W0055365

Fortschritte der Chemie organischer Naturstoffe

Progress in the Chemistry of Organic Natural Products

64

Founded by L. Zechmeister
Edited by W. Herz, G. W. Kirby, R. E. Moore,
W. Steglich, and Ch. Tamm

Authors:
J. Bermejo Barrera, H. J. M. Gijsen, A. G. González,
Ae. de Groot, G. Prota, J. B. P. A. Wijnberg

Springer-Verlag
Wien New York 1995

Prof. W. HERZ, Department of Chemistry,
The Florida State University, Tallahassee, Florida, U.S.A.

Prof. G. W. KIRBY, Chemistry Department,
The University, Glasgow, Scotland

Prof. R. E. MOORE, Department of Chemistry,
University of Hawaii at Manoa, Honolulu, Hawaii, U.S.A.

Prof. Dr. W. STEGLICH, Institut für Organische Chemie der Universität
München, München, Federal Republic of Germany

Prof. Dr. CH. TAMM, Institut für Organische Chemie der Universität Basel,
Basel, Switzerland

This work is subject to copyright.
All rights are reserved, whether the whole or part of the material is concerned, specifically those
of translation, reprinting, re-use of illustrations, broadcasting, reproduction by photocopying
machines or similar means, and storage in data banks.

© 1995 by Springer-Verlag/Wien
Softcover reprint of the hardcover 1st edition 1995

Library of Congress Catalog Card Number AC 39-1015

Typesetting: Macmillan India Ltd., Bangalore-25

Printed on acid free and chlorine free bleached paper

With 22 Figures, partly coloured

ISSN 0071-7886
ISBN-13:978-3-7091-9339-6   e-ISBN-13:978-3-7091-9337-2
DOI: 10.1007/978-3-7091-9337-2

Contents

List of Contributors

BERMEJO BARRERA, Prof. J., Centro de Productos Naturales Orgánicos "A. González", Instituto de Productos Naturales y Agrobiologia de Canarias C.S.I.C., Avenida Astrofísico Fco. Sánchez 2, E-38206 La Laguna, Tenerife, Canary Islands, Spain.

GIJSEN, Dr. H. J. M., Laboratory of Organic Chemistry, Wageningen Agricultural University, Dreijenplein 8, NL-6703 HB Wageningen, The Netherlands.

GONZÁLEZ, Dr. A. G., Centro de Productos Naturales Orgánicos "A. González", Instituto de Productos Naturales y Agrobiologia de Canarias C. S. I. C., Avenida Astrofísico Fco. Sánchez 2, E-38206 La Laguna, Tenerife, Canary Islands, Spain.

GROOT, Prof. Dr. AE. DE, Laboratory of Organic Chemistry, Wageningen Agricultural University, Dreijenplein 8, NL-6703 HB Wageningen, The Netherlands.

PROTA, Prof. Dr. G., Department of Organic and Biological Chemistry, University of Naples, Via Mezzocannone 16, I-80134 Naples, Italy.

WIJNBERG, Dr. J. B. P. A., Laboratory of Organic Chemistry, Wageningen Agricultural University, Dreijenplein 8, NL-6703 HB Wageningen, The Netherlands.

Chemistry and Sources of Mono- and Bicyclic Sesquiterpenes from *Ferula* Species

A. G. González and J. Bermejo Barrera, Centro de Productos Naturales Orgánicos "A. González", Instituto de Productos Naturales y Agrobiología de Canarias C.S.I.C., La Laguna, Tenerife, Canary Islands, Spain

Contents

I. Introduction

The exclusively Old World genus *Ferula* belongs to the family Umbelliferae (tribe Peucedaneae, subtribe Peucedaninae) (*1*) with some 130 species distributed throughout the Mediterranean area and Central Asia. The highest concentration of species is in the former USSR and neighbouring countries where nearly 100 species have been described, about 60 of them endemic. More than twenty species have been reported in Iran and three in the Canary Islands off the Atlantic coast of Africa (Figs. 1, 2 and 3).

A review of the chemical constituents of *Ferula* was published by SAIDKHODZHAEV in 1979 who attributed the first serious chemical studies in this field to the group headed by I. P. TUKERVANKIK *et al.* in 1935. These researchers published several papers on characteristic *Ferula* resins (*2*).

A. G. GONZÁLEZ and J. BERMEJO BARRERA

Fig. 1. *Ferula linkii-GC*
Fig. 2. *Ferula latipinna*
Fig. 3. *Ferula lancerottensis*

References, pp. 82–92

By the early sixties N. P. KIR'YALOV *et al.* were studying the composition of *Ferula* as a good source of biologically active compounds (*3–8*) and were the first to establish that *Ferula* species contained sesquiterpene lactones and alcohols as well as coumarins (*9, 10*).

Essential oils and coumarins have been the primary subject of chemical research on *Ferula* species and terpenoid coumarins, aromatic acid esters, terpene alcohols, sesquiterpene lactones (*11*) and secondary propenyl bisulphides (*12*) have all been reported as constituents of roots and gum resins. Several species of *Ferula* have been used in folk medicine; thus, *Ferula communis* L. and its subspecies and varieties have been used as agents against hysteria and to treat dysentery (*13*), *F. jaeschkeana* Vatke has been applied to wounds and bruises (*14*) and *F. tingitana* L. has proved to be a good source of ammoniac, an oleo-gum resin used in medicine (*15*). The study of the chemistry of the genus *Ferula* has developed rapidly over the last twenty years due to more efficient methods of purification and the availability of increasingly sophisticated techniques for structure elucidation.

Although a short review of metabolites from *Ferula* species was published in 1988 (*16*), it was far from exhaustive. The present article is intended to provide a complete account of mono- and bicyclic sesquiterpene derivatives obtained from plants of the genus *Ferula* during the period 1970–1990 as well as details of isolation techniques and the structural determination, synthesis and stereochemistry of some of the more representative species.

II. Biosynthetic Relationships

The sesquiterpenes are a very large group of natural hydrocarbons of composition $C_{15}H_{24}$ and are the most extensive terpene group, the term "terpene" being understood to mean compounds which have a chemical and structural relationship with the basic isoprene molecule (*17*).

WALLACH suggested a general formula for the sesquiterpenes based on the idea that they are composed of three isoprene molecules; if these three molecules are adequately combined, structure (**1**) is obtained and from a structural point of view may be considered to be derived from substituted and partially hydrogenated naphthalene. Although it is now recognised that the isoprene nuclei were not put in the right order, Wallach's idea is still of the greatest importance as almost all known sesquiterpenes do contain three isoprene molecules. Later it was realised that a similar type of unit construction could be applied to more complex molecules and this

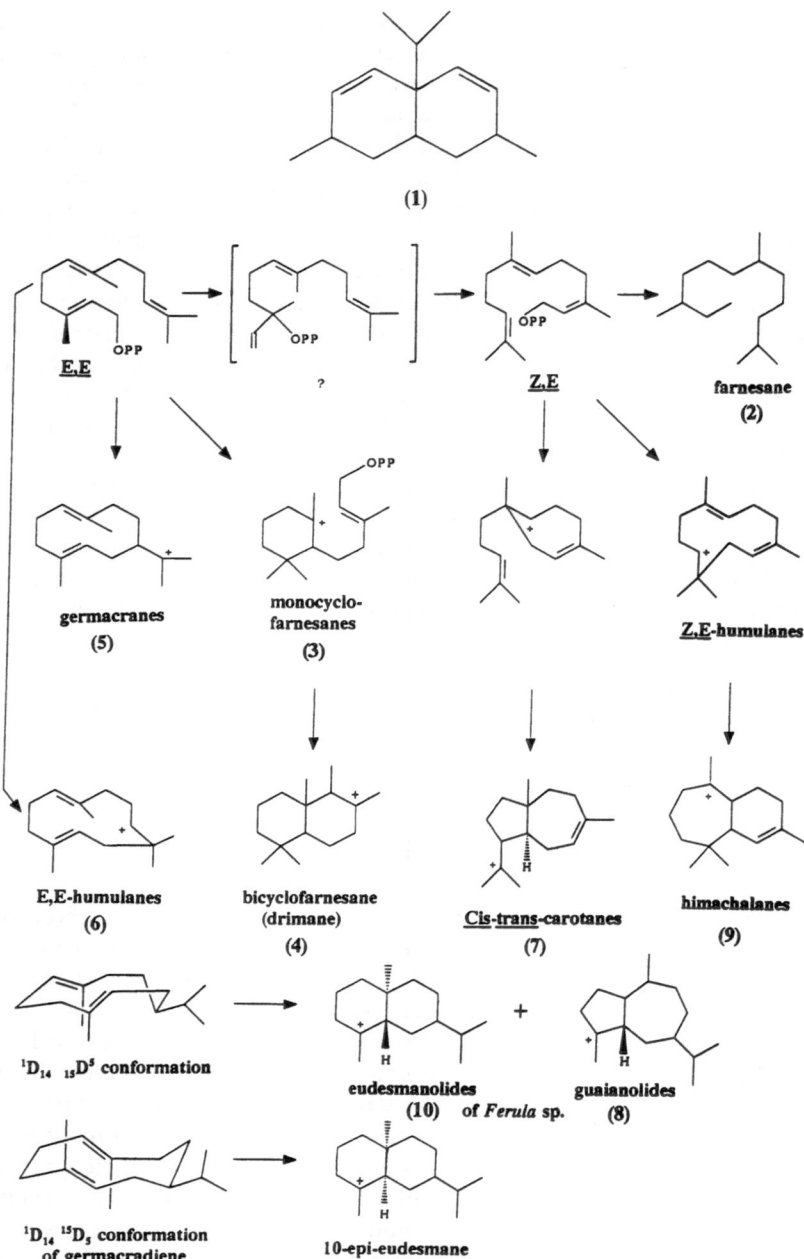

Chart 1. Biogenetic sesquiterpene interconversions in *Ferula*

led to RUZICKA's formulation of the biogenetic isoprene rule (*18*) as later ratified by HENDRICKSON (*19*).

Once this rule was understood, efforts were made to discover which biological processes could take place in vegetable tissue to biosynthesize these substances. Nowadays, experiments have shown (*20*) that the critical point in the biogenesis of sesquiterpenes is the formation of 2,3-*cis*- or *trans*-farnesol, in which three isoprene units are linked head-to-tail.

To date derivatives of nine types of sesquiterpenes have been isolated from plants of the genus *Ferula*; farnesane (**2**), monocyclofarnesane (**3**), bicyclofarnesane (**4**), germacrane (**5**), humulane (**6**), carotane (**7**), guaiane (**8**), himachalane (**9**) and eudesmane (**10**). Sesquiterpene esters are much less homogeneous than lactones and since five types (**5–9**) have already been found (Chart 1) it is reasonable to assume that other sesquiterpene types will also be identified, eventually.

III. Extraction and Isolation

Plants are generally dried, ground and then extracted in a Soxhlet or left to macerate at room temperature. The extracts are evaporated under reduced pressure giving material which is then extracted with methanol, ethanol or benzene. After removal of solvent, the residue is dissolved in a mixture of methanol-water. The soluble part is extracted with petrol ether or chloroform, yielding triterpenoids, sterols and fatty acids and the insoluble part chromatographed to afford the sesquiterpenes (Chart 2).

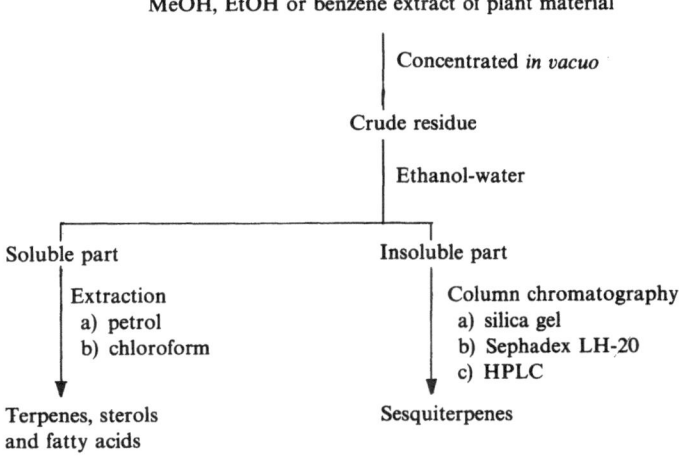

Chart 2. Isolation procedure for sesquiterpene derivatives

IV. Determination and Structural Types

Sesquiterpene derivatives have been isolated from several species of *Ferula*. KIR'YALOV *et al.* (*10*) were the first to show that the genus *Ferula* not only contained coumarins and sesquiterpene lactones, as mentioned in the Introduction, but also sesquiterpenes esterified by aromatic and aliphatic acids. These sesquiterpenes are divided into monocyclic and bicyclic compounds according to their carbon skeleton. The monocyclic sesquiterpene alcohols are germacranes or humulanes and the bicyclic sesquiterpene alcohols are of the carotane, guaiane, himachalane, selinane and camphane types (Chart 3). These alcohols are esterified by vanillic, isovanillic, veratric, *p*-hydroxybenzoic, methoxybenzoic, 3,4-dihydroxybenzoic, angelic, tiglic and acetic acids.

Physical methods are of great use in the structure determination of these compounds: infrared, ultraviolet and mass spectrometry and nuclear magnetic resonance (¹H and ¹³C) have revolutionised natural product chemistry while X-ray studies can clear up specific structural problems. Two absorption regions in the IR spectra are significant for the chemistry of these products: that between 1520–1620 cm^{-1} (benzene ring) shows if the substance is the ester of an aromatic acid, and that between 1690–1710 cm^{-1}, if intense, is characteristic of the presence of carbonyl

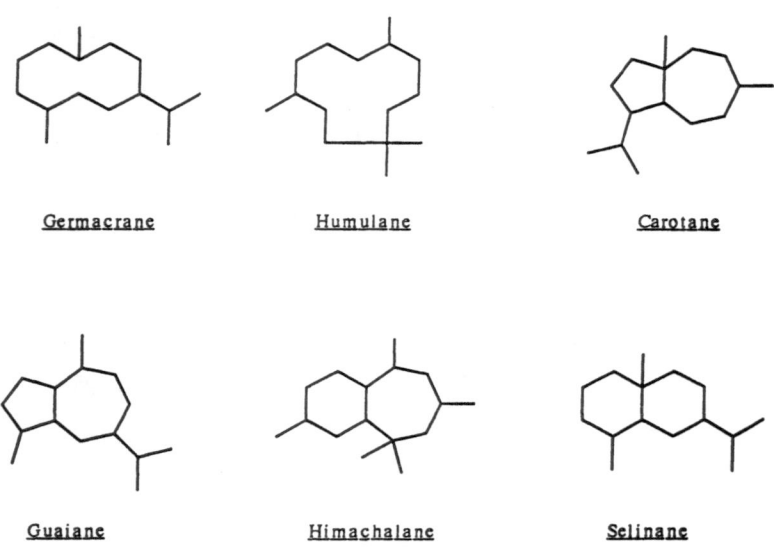

Germacrane   Humulane   Carotane

Guaiane   Himachalane   Selinane

Chart 3. Structural types of sesquiterpenes from *Ferula*

groups (ester carbonyl). Ultraviolet spectroscopy is principally used to determine the presence or absence of conjugation. WOODWARD's rules modified by FIESER (*21*) are valuable in forecasting the positions of the maxima of conjugated dienes, trienes and ketones. Nuclear magnetic resonance (^{1}H and ^{13}C) and mass spectrometry also provide basic information about the structures. In addition to these physical techniques, correlation with fully-characterized substances is the best method of all for total elucidation of the structure of new compounds.

V. Monocyclic Sesquiterpenes

The earliest Soviet researchers gave only provisional structures for a certain number of sesquiterpenes; hence many have since had to be altered. In 1978, for instance, GOLOVINA *et al.* (*22*) isolated five germacrane sesquiterpenes tschimganidin, ferolin, federin, rubaferinin and rubaferidin from the ethanol extract of the roots of *F. rubroarenosa* which on the basis of the physicochemical constants of the alkaline hydrolysis products, spectral characteristics and chemical degradations were assigned structures (**11a–15a**) shown below.

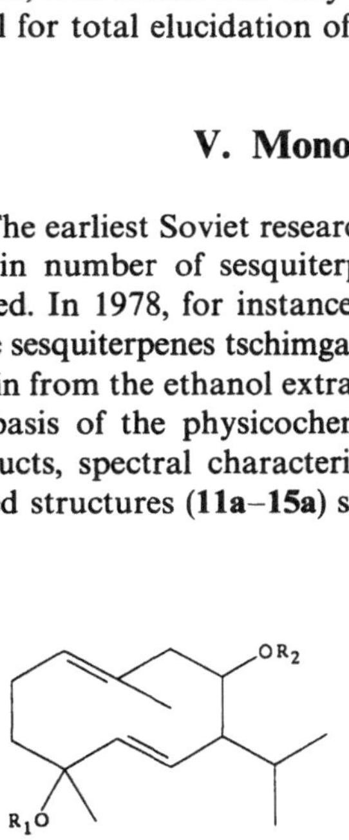

(11) R_1 = H, R_2 = Van
(12) R_1 = H, R_2 = p-HyBz
(13) R_1 = Ac, R_2 ≡ p-HyBz

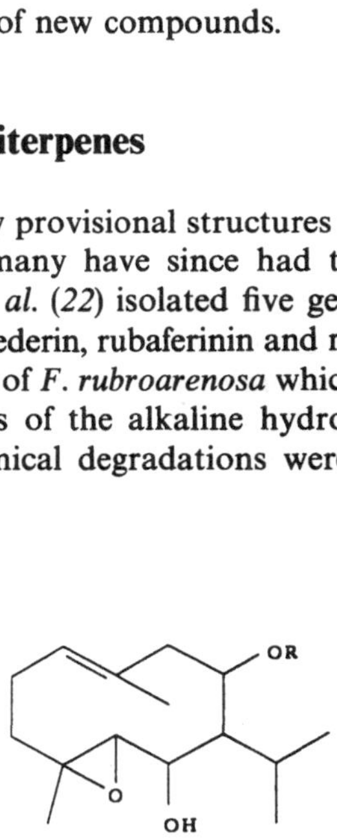

(14)   R = Van
(15)   R = p-HyBz

Tschimganidin, ferulin, rubaferinin and rubaferidin (actual structures **11, 12, 14** and **15**) were later isolated from *F. orientalis* var. *orientalis* (*23*), and federin (actual structure **13**) from *F. ovina* (*24*). In all cases their structures were modified and their stereochemistry was assigned on the basis of their spectral (principally ^{1}H NMR) data. Full details of the spectra of (**11**) and (**12**) (*23*) are given here as the initial Soviet reports were based on limited data. These two compounds, (**11**) and (**12**), were later given the names 8-*p*-vanilloyltovarol and 8-*p*-hydroxybenzoyl-tovarol.

(11)	R_1 = H, R_2 = Van
(12)	R_1 = H, R_2 = p-HyBz
(13)	R_1 = Ac, R_2 = p-HyBz

(14)	R = Van
(15)	R = p-HyBz

As can be seen from Tables 1 and 2, it is unusual for daucanes and germacranes to co-occur in one species (*25, 26*). Apart from *F. orientalis* var. *orientalis*, such compounds are only found together in *F. rubroarenosa (22)* and *F. tenuisecta* Korovin (*27*). *F. orientalis, F. rubroarenosa* and *F. tenuisecta* have been placed in the *Ovina* complex of section *Xeronarthex* of *Ferula* subgenus *Peucedanoides*, by Korovin (*28*) and the association of these particular sesquiterpene metabolites in the three species supports this taxonomical classification.

Miski *et al.* (*29*) isolated ten new sesquiterpenes from *Ferula haussknechtii* and elucidated their structures by a combination of spectroscopic methods (2D NMR homo- and heteronuclear COSY and high resolution NOESY experiments) and chemical degradations. Substances (**16**), (**17**) and (**18**) proved to be fervanol benzoate, p-hydroxybenzoate and vanillate, respectively, and all had IR absorptions characteristic of an acyl aromatic group.

The ^{1}H NMR spectra of (**16–18**) indicated the presence of an exocyclic methylene group, a trisubstituted double bond and a *trans* disubstituted double bond as well as two geminal methyl groups. The presence of these three double bonds indicated a monocyclic structure for all three compounds; moreover, the presence of two geminal methyl groups instead of an isopropyl group clearly demonstrated a γ-humulenoid-type structure for (**16–18**). In the ^{1}H NMR spectra of these substances the exocyclic methylene proton signals appeared as two doublets, whereas the same signals had been reported as broad singlets in the spectra of all previously known γ-humulenes except for the compound obtained from *F. ceratophylla* which probably has the same stereochemical features as (**16–18**) (*30, 31*). Nevertheless, the spectral data did not establish the full stereochemistry which was later ascertained by NOESY experiments.

The location of the aromatic acyl groups and the structural relationship of these three compounds were deduced from 2D NMR homonuclear COSY-45° experiments. The 2D NMR homonuclear COSY-45° experiment confirmed the presence of geminal coupling between the exocyclic methylene protons but no other long-range couplings with these protons, suggesting that (16–18) had a conformation different from that of the previously reported γ-humulenes. Further information regarding the stereochemistry of these substances was obtained by a series of NOESY experiments. The ^{1}H NMR NOESY experiment on (18) confirmed the S-*trans* configuration of the conjugated 4(15),5-diene system and the β orientation of the C-7 acyl group.

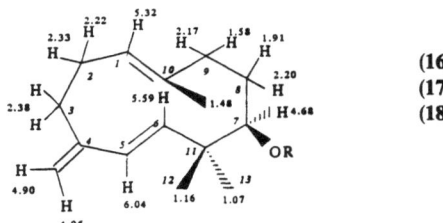

(16) R = Bz
(17) R = p-HyBz
(18) R = Van

Kurubasch aldehyde benzoate (19) and vanillate (20), the name deriving from the collection site, had IR absorption bands characteristic of an α, β-unsaturated aldehyde and an aromatic acyl group. The aromatic acyl groups of the two substances were identified as benzoate and vanillate, respectively, from their ^{1}H, ^{13}C NMR and MS data.

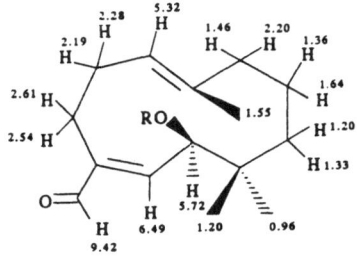

(19) R = Bz

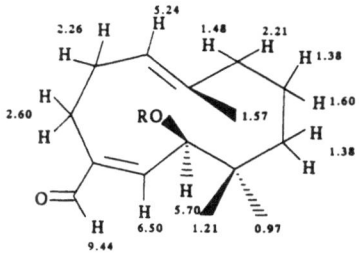

(20) R = Van

The similarity of the spectra, except for the nature of the acyl group, confirmed that both substances possessed the same sesquiterpene skeleton. It could be deduced from the results of a 2D NMR homonuclear COSY-45° experiment that the structures of **19** and **20** were of the γ-humulene type but the downfield ^{1}H NMR position of the H-5 signals (δ 6.50) and the presence of an aldehyde proton signal (δ 9.40) instead of the second vinyl methyl signal to be expected in a γ-humulene, clearly indicated that in (**19**) and (**20**) the C-4 methyl of the γ-humulenes had been replaced by an aldehyde group. The doublet at δ 5.70 which was coupled with the H-5 signal at δ 6.50 confirmed the location of the aromatic acyl groups at C-6 in both compounds. NOESY experiments applied to (**20**) revealed the E configuration of the $\Delta^{1,10}$ and Δ^4 double bonds and the β stereochemistry of the acyl group at C-6.

Kurubasch acid angelate (**21**) and benzoate (**22**) displayed spectral data similar to those of (**19**) and (**20**). Thus the ^{1}H NMR spectrum of (**22**) was only slightly different from that of (**19**), one of these differences being the absence of an aldehyde. The IR, ^{13}C NMR and mass spectra of (**22**) confirmed that the C-4 aldehyde group of (**19**) was oxidized to a carboxy group in (**22**).

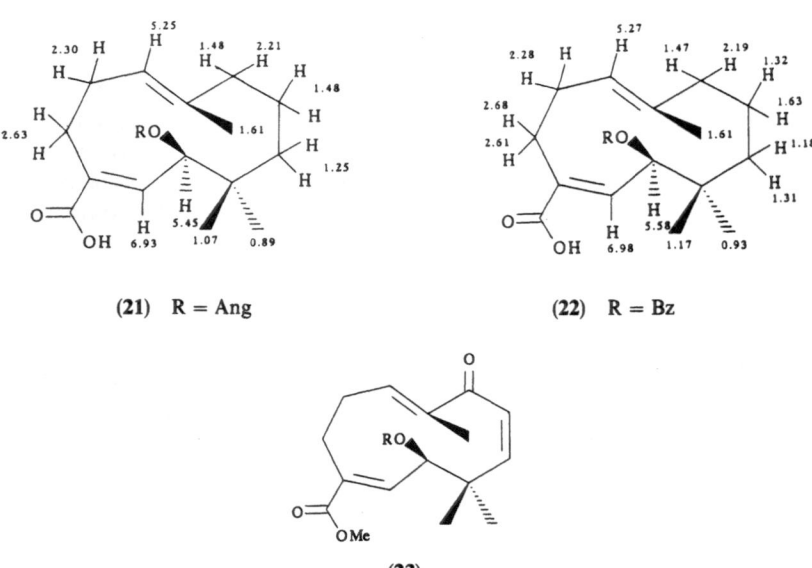

(21)   R = Ang   **(22)   R = Bz**

(23)

The similar chemical shifts of H-5 in the ^{1}H NMR spectra of (**22**) and (**23**), the latter prepared earlier by UV irradiation (*32*) of the 4Z isomer suggested that (**22**) also had the 4*E* configuration and this was borne out when (**19**) was oxidized with chromic acid to form (**22**). The spectral data of (**24**), (**25**) and (**26**) indicated that they were all 1,10-epoxy analogues of kurubasch acid esters, with (**24**) and (**25**) possessing the same stereo-chemistry and (**26**) being different.

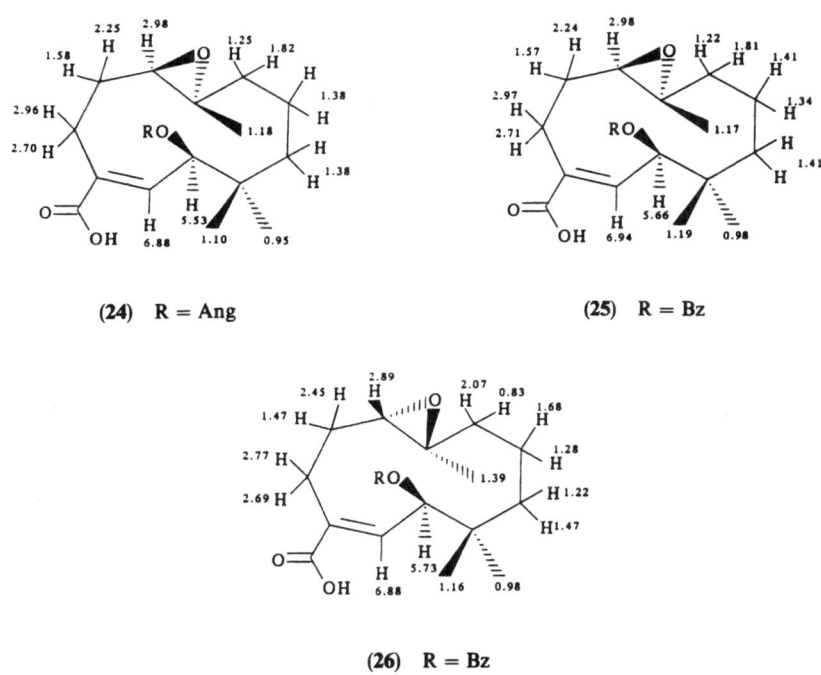

(**24**)   R = Ang   (**25**)   R = Bz

(**26**)   R = Bz

^{1}H NMR spectra (*29*)

GONZALEZ *et al.* (*33*) obtained three new humulene sesquiterpenes (**56**), (**57**) and (**58**), from *F. latipinna* Santos (Fig. 2) which exhibited IR absorptions for an aromatic acyl group identified as a veratrate from the ^{1}H, ^{13}C NMR and MS data.

The ^{1}H NMR spectrum of (**56**) is similar to that of (**19**), (**20**), (**21**) and (**22**) but exhibits two vinyl methyl signals. Location of the aromatic acyl group at C-6 was deduced from the coupling of the H-6 doublet at δ 5.50 to the H-5 signal at δ 5.40. NOE experiments on (**56**) showed that the $\Delta^{1,10}$ and Δ^4 double bonds had, respectively, *E* and *Z* configurations, and that the stereochemistry of the acyl group at C-6 was β (*29*). Compounds (**57**) and (**58**) and ^{1}H and ^{13}C NMR spectra analogous to those of (**56**),

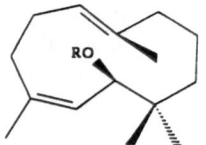

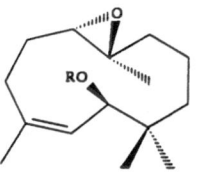

(56) R = Ver

(30) R = HL
(31) R = p-Br-C₆H₄CO
(57) R = VeR

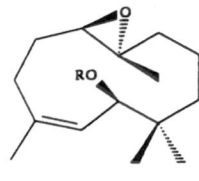

(58) R = Ver

with the addition of a 1,10-oxirane ring. The disposition of this ring was determined by spectroscopic analysis.

Alkaline hydrolysis of (57) gave veratric acid and a monodeacyl derivative (30) analogous to (24), (25) and (26). The p-bromobenzoyl ester (31) of (57) was identical with one described earlier by ITOKAWA et al. (34). Compound (58) proved to be analogous to (57), the only difference being the epoxide stereochemistry.

Humulenoid sesquiterpenes are relatively rare in nature and the cis Δ^4 configuration of the compounds described above distinguishes them from the more common all-trans-humulene type. One may speculate that compounds with a cis Δ^4-double bond (α-apienes) are cyclization products of cis, trans-farnesyl pyrophosphate (19) and not of trans,trans-farnesyl pyrophosphate which is the precursor of the more common trans-humulenes. It is noteworthy that to date all humulenoids with a cis Δ^4-double bond or S-trans-4(15),5 dienes have been isolated from the Apiaceae; for this reason they have been placed in a subclass known as apienes in a manner similar to the germacrene-heliangene relationship. The distribution and structures of all monocyclic sesquiterpenes from Ferula species are shown in Table 1 and Chart 4, respectively. In a number of cases the stereochemistry remains to be determined.

Table 1. *Monocyclic Sesquiterpenes Found in Ferula*

Structure number	Name of compound	Formula	Plant source	References
(11)	Tschimganidin	$C_{23}H_{32}O_5$	*Ferula rubroarenosa*	(22)
			F. orientalis var. *orientalis*	(23)
			F. angrenii	(94)
			F. calcarea	(95)
			F. kopetdaghensis	(96)
			F. lapidosa	(97)
			F. pallida	(98)
			F. tenuisecta	(99)
			F. tschimganica	(100)
(12)	Ferolin	$C_{22}H_{30}O_4$	*F. rubroarenosa*	(22)
			F. orientalis var. *orientalis*	(23)
			F. angrenii	(94)
			F. calcarea	(95)
			F. kopetdaghensis	(96)
			F. lapidosa	(97)
			F. pallida	(98)
			F. tenuisecta	(99)
			F. tschimganica	(100)
(13)	Federin	$C_{24}O_5$	*F. rubroarenosa*	(22)
			F. fedtschenkoana	(101)
			F. ovina	(24)
(14)	Rubaferinin	$C_{23}H_{32}O_6$	*F. rubroarenosa*	(22)
			F. orientalis var. *orientalis*	(23)
			F. calcarea	(95)
			F. kopetdaghensis	(96)
			F. tenuisecta	(102)

Table 1 (*continued*)

Structure number	Name of compound	Formula	Plant source	References
(15)	Rubaferidin	$C_{22}H_{30}O_5$	*F. rubroarenosa*	(22)
			F. calcarea	(95)
			F. kopetdaghensis	(96)
			F. orientalis var. *orientalis*	(23)
			F. tenuisecta	(102)
(16)	Fervanol benzoate	$C_{22}H_{28}O_2$	*F. haussknechtii*	(29)
(17)	Fervanol 1-hydroxybenzoate	$C_{23}H_{30}O_4$	*F. haussknechtii*	(29)
(18)	Fervanol vanillate	$C_{23}H_{30}O_4$	*F. haussknechtii*	(29)
(19)	Kurubasch aldehyde benzoate	$C_{22}H_{28}O_3$	*F. haussknechtii*	(29)
(20)	Kurubasch aldehyde vanillate	$C_{23}H_{30}O_5$	*F. haussknechtii*	(29)
(21)	Kurubasch acid angelate	$C_{20}H_{30}O_4$	*F. haussknechtii*	(29)
(22)	Kurubasch acid benzoate	$C_{22}H_{28}O_4$	*F. haussknechtii*	(29)
(24)	1β,10α-Epoxykurubasch acid angelate	$C_{20}H_{30}O_5$	*F. haussknechtii*	(29)
(25)	1β,10α-Epoxykurubasch acid benzoate	$C_{22}H_{28}O_5$	*F. haussknechtii*	(29)
(26)	1α,10β-Epoxykurubasch acid benzoate	$C_{22}H_{28}O_5$	*F. haussknechtii*	(29)
(27)	Ferocin	$C_{23}H_{32}O_3$	*F. ceratophylla*	(103)
			F. helenae	(102)
			F. lapidosa	(102)
			F. subtilis	(104)
			F. tenuisecta	(102)
(28)	Ferocinin	$C_{24}H_{34}O_4$	*F. ceratophylla*	(103)
			F. helenae	(102)
			F. lapidosa	(102)
			F. subtilis	(104)

No.	Compound	Formula	Species	Ref.
(29)	Ferocidin = Juperin (32)	$C_{22}H_{28}O_3$	*F. temuisecta*	(102)
(32)	Juperin = Ferocidin (29)	$C_{22}H_{28}O_3$	*F. tschatcalensis*	(105)
(33)	Fekserol	$C_{15}H_{26}O_3$	*F. ceratophylla*	(103)
			F. ceratophylla	(107)
			F. juniperina	(108)
(34)	Juniperin	$C_{23}H_{32}O_5$	*F. tschatcalensis*	(106)
			F. ceratophylla	(107)
			F. juniperina	(108)
(35)	Juniperinin	$C_{22}H_{30}O_4$	*F. tschatcalensis*	(105)
			F. ceratophylla	(107)
			F. juniperina	(108)
(36)	Juniperidin	$C_{24}H_{32}O_5$	*F. ceratophylla*	(107)
			F. juniperina	(108)
(37)	Allokedycariol	$C_{15}H_{26}O$	*F. communis* var. *communis*	(64)
(38)	Federin	$C_{24}H_{32}O_5$	*F. fedtschenkoana*	(101)
			F. ovina	(24)
(39)	Angrendiol = Fetinin (41)	$C_{22}H_{30}O_4$	*F. tschimganica*	(109)
(40)	Ugamdiol trimethoxy-gallicate = Ugaferin (45)	$C_{25}H_{36}O_7$	*F. leucographa*	(112,113)
(41)	Fetinin = Angrendiol (39)	$C_{22}H_{30}O_4$	*F. temuisecta*	(115)
(42)	Fecorine	$C_{22}H_{34}O_5$	*F. korshinskyi*	(110)
(43)	Involucrin	$C_{27}H_{38}O_8$	*F. involucrata*	(111)
			F. leucographa	(112)
			F. ugamica	(112)
(44)	Involucrinin	$C_{30}H_{42}O_8$	*F. involucrata*	(111)
			F. leucographa	(112)
			F. ugamica	(112)
(45)	Ugaferin = (40)	$C_{25}H_{34}O_7$	*F. involucrata*	(111)
			F. leucographa	(112)
			F. ugamica	(112)

Table 1 (continued)

Structure number	Name of compound	Formula	Plant source	References
(46)	6-β-p-Hydroxybenzoyloxy-germacra-1(10),4-diene	$C_{22}H_{30}O_3$	F. orientalis var. orientalis	(24)
(47)	6-β-Vanilloyloxy-germacra-1(10),4-diene	$C_{23}H_{32}O_5$	F. orientalis var. orientalis	(24)
(48)	Fertenin	$C_{22}H_{30}O_5$	F. tenuisecta	(114)
(49)	Fertenidine	$C_{22}H_{32}O_5$	F. tenuisecta	(27)
(50)	Fertenicine	$C_{22}H_{30}O_4$	F. tenuisecta	(27)
(51)	Fexerin	$C_{20}H_{32}O_3$	F. tschatcalensis	(105)
(52)	Fekserinin	$C_{23}H_{32}O_4$	F. xeromorpha	(116)
(53)	Fekserine	$C_{20}H_{32}O_3$	F. xeromorpha	(116)
(54)	Chafterin	$C_{20}H_{32}O_4$	F. tschatcalensis	(105)
(55)	Fekseridin	$C_{23}H_{34}O_6$	F. xeromorpha	(117)
(56)	7,8-Dihydro-α-humulene 6β-veratrate	$C_{24}H_{34}O_4$	F. latipinna	(33)
(57)	1α,10β-Epoxy-7,8-dihydro-α-humulene 6β-veratrate	$C_{24}H_{34}O_5$	F. latipinna	(33)
(58)	1β,10α-Epoxy-7,8-dihydro-α-humulene 6β-veratrate	$C_{24}H_{34}O_5$	F. latipinna	(33)

Chart 4. Structures of monocyclic sesquiterpenes found in *Ferula*

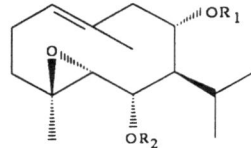

(11) Tschimganidin (Tovarol 8-vanillate); R_1 = Van, R_2 = H
(12) Ferolin (Tovarol 8-*p*-hydroxybenzoyl); R_1 = *p*-HyBz, R_2 = H
(13) Federin; R_1 = *p*-HyBz, R_2 = Ac

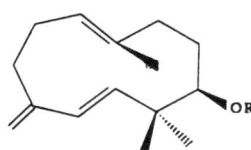

(14) Rubaferinin (Shiromodiol 8-vanillate); R_1 = Van, R_2 = H
(15) Rubaferidin (Shiromodiol 8-*p*-hydroxybenzoate); R_1 = *p*-HyBz, R_2 = H

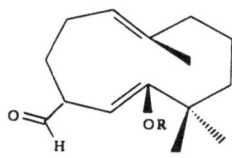

(16) Fervanol benzoate; R = Bz
(17) Fervanol *p*-hydroxybenzoate; R = *p*-HyBz
(18) Fervanol vanillate; R = Van

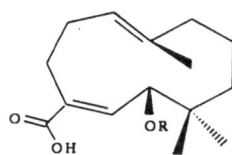

(19) Kurubasch aldehyde benzoate; R = Bz
(20) Kurubasch aldehyde vanillate; R = Van

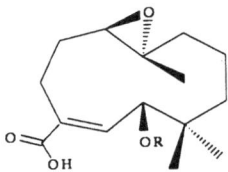

(21) Kurubasch acid angelate; R = Ang
(22) Kurubasch acid benzoate; R = Bz

(24) 1β,10α-Epoxykurubasch acid angelate; R = Ang
(25) 1β,10α-Epoxykurubasch acid benzoate; R = Bz

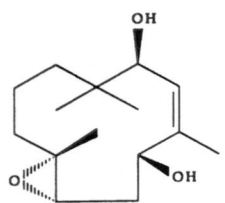

(26) 1α,10β-Epoxykurubasch acid benzoate; R = Bz

(27) Ferocin; R = *p*-HyBz
(28) Ferocinin; R = Van

(29) Ferocidin; R = *p*-HyBz
(32) Juperin; R = *p*-HyBz

(33) Fekserol

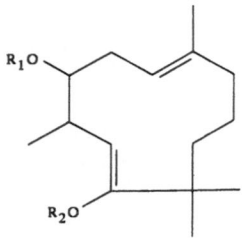

(34) Juniperin; R₁ = Van, R₂ = H
(35) Juniperin; R₁ = *p*-HyBz, R₂ = H
(36) Juniperidin; R₁ = p-HyBz, R₂ = Ac

(37)   Allokedycariol

(38)   Federin; R_1 = p-HyBz, R_2 = Ac
(39)   Angrendiol; R_1 = p-HyBz, R_2 = H
(40)   Ugamdiol trimethoxy gallicate; R_1 = 3,4,5-Tri-
           methoxybenzoic acid, R_2 = H
(41)   Fetinin; R_1 = p-HyBz, R_2 = H

(42)   Fecorine; R_1 = Ac, R_2 = Ang

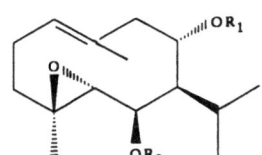

(43)   Involucrin; R_1 = 3,4,5-Trimethoxybenzoic acid,
           R_2 = Ac
(44)   Involucrinin; R_1 = 3,4,5-Trimethoxybenzoic
           acid, R_2 = Ang
(45)   Ugaferin; R_1 = 3,4,5-Trimethoxybenzoic acid,
           R_2 = H

(46)   6β-p-Hydroxybenzoyloxygermacra-1(10),4-
           diene; R = p-HyBz
(47)   6β-Vanilloyloxygermacra-1(10),4-diene;
           R = Van

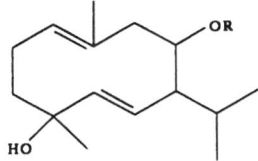

(48)   Fetinin; R = p-HyBz

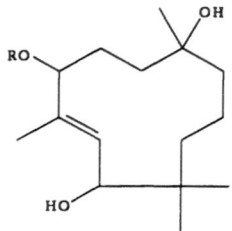

(49)   Fertenidin; R = p-HyBz

(50)   Fertenicin; R = p-HyBz

(51)   Fexerin; R_1 = OH, R_2 = Ang
(52)   Fecserinin; R_1 = H, R_2 = iVan
(53)   Fekserine; R_1 = OH; R_2 = Tig

(54)   Chatferin; R = Tig

(55)   Fecseridin; R = iVan

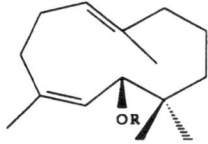

(56) 7,8-Dihydro-α-humulene 6β-veratrate; R = Ver

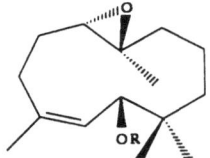

(57) 1α,10β-Epoxy-7,8-dihydro-α-humulene 6β-veratrate; R = Ver

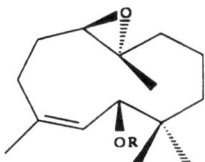

(58) 1β,10α-Epoxy-7,8-dihydro-α-humulene 6β-veratrate; R = Ver

VI. Bicyclic Sesquiterpenes

As stated earlier (Chart 3), bicyclic sesquiterpenes from *Ferula* species fall into the carotane, guaiane, himachalane, selinane and camphane categories, with carotane-type skeletons being found most frequently.

In 1976, an investigation of galbanum resin from *Ferula galba galbanifula* (*35*) yielded the selinane, 10-epijunenol (**59**), a new and unusual natural *cis*-eudesmane.

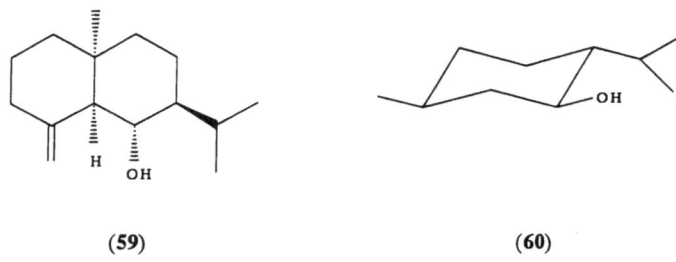

(59)   **(60)**

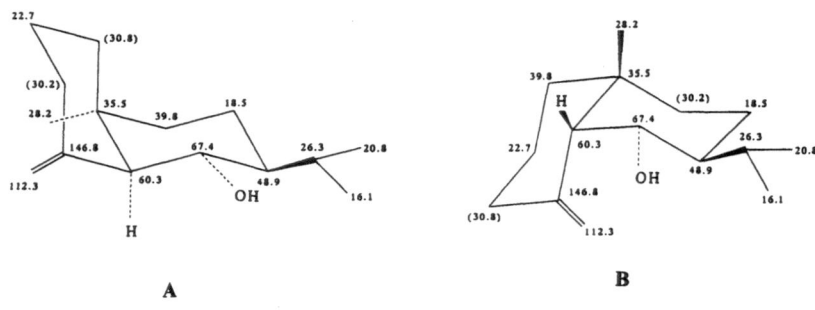

Chart 5. ^{13}C NMR data (*35*)

The ^{1}H NMR spectrum of (**59**) indicated the presence of a carbinol proton, a saturated isopropyl group, a tertiary methyl and a methylene group, while the ^{13}C NMR spectrum clearly showed that the molecule was *cis*-fused and that there were two structures to be considered (Chart 5). The values in parentheses are alternative attributions.

The chemical shift of the methine carbon to which the isopropyl group was attached favoured structure **A**. In this case, the chemical shift of the C-7 (48.9 ppm) should be close to that of C-4 (50.0 ppm) in menthol (**60**) (*36*). In structure **B** the chemical shift would be expected to be about 5–6 ppm upfield from C-4 of (**60**) as a result of steric compression exercised by the C-5 axial substituent in the *cis*-fused system. The chemical shift of the methylene carbon adjacent to the double bond also supported structure **A**. This structure was further corroborated when the ^{1}H NMR spectrum was compared with those of the menthols (*37*).

The structure of 10-epijunenol (**59**) was confirmed by the following chemical transformations: (**59**), when oxidized with CrO$_3$ (*38*), gave a single ketone (**61**), and unchanged epijunenol (**59**). Base-catalyzed isomer-ization of (**61**) with sodium methoxide in methanol gave a mixture containing unchanged (**61**), isomer (**62**) and two isomers of a conjugated ketone, (**63**) and (**64**) (Chart 6).

The stereochemistry of (**61**) was apparent from the unchanged posi-tion of the ^{1}H NMR signal of the angular methyl group which would be at higher field in a *trans*-10-methyldecalone. Thus Asakawa *et al.* (*39*) reported 0.82 ppm for the chemical shift of H-14 in a substituted (*trans*) junenol while in epijunenol (**59**) it appeared at 0.89 ppm. Furthermore, it has been shown (*40*) that one of the methylene protons of a γ-methylene-*trans*-10-methyldecal-6-one suffers a large anisotropic shift owing to the proximity of the carbonyl group so that the signals of the methylene group resemble those of a 2-methylenecyclohexanone. In (**61**) and (**62**)

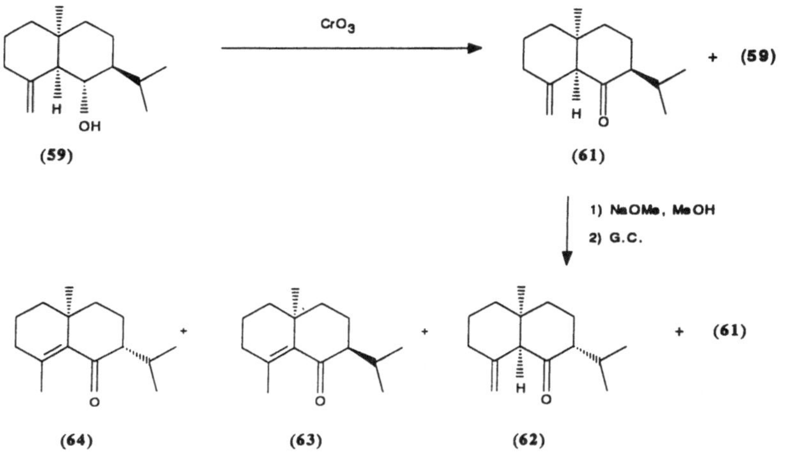

Chart 6. The chemical transformation of epijunenol (59)

there is only a minor displacement from the corresponding signals of epijunenol (59), demonstrating that the carbonyl group is farther away in *cis* isomers. The conjugated ketones (63) and (64) have been described (*40*) and their constitutions have been further corroborated by an independent synthesis starting from octalone (65) (*41*). Ethoxycarboxylation of (65) yielded an enolized β-ketoester, (66), with traces of the keto form (67) which, on treatment with excess methyl lithium, was transformed into the hydroxyketone (68). Elimination of water (thionyl chloride in pyridine) gave the unstable isopropenyl ketone (69) which was reduced with

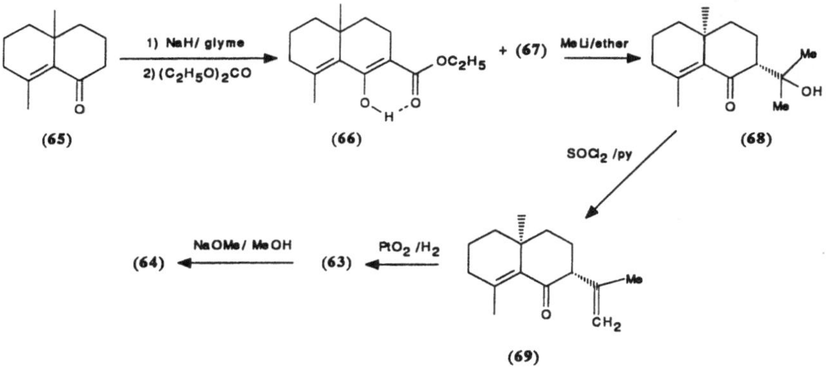

Chart 7. Synthesis of selinanes (63) and (64)

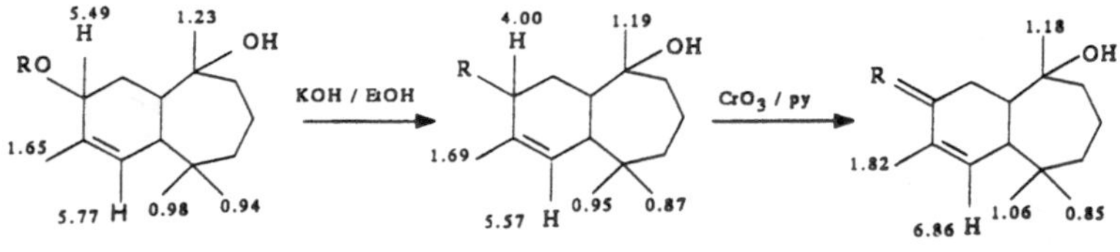

Chart 8. Reactions and spectral data of xeroferin (70)

hydrogen on platinum to the isopropyl ketone, (63), identical in all respects with the ketone obtained from the natural product. Isomerization of ketone (63) yielded a mixture containing the stereoisomer (64) as a minor product (Chart 7).

From the roots of *Ferula xeromorpha*, BIZHANOVA *et al.* isolated a himachalane ester [xeroferin (70)] for the first time from this genus (42). The gross structure of (70) was deduced from its ^{1}H NMR spectrum and chemical characteristics (Chart 8). Treatment of (70) with KOH-EtOH gave a diol (71) (xeroferol) and isovanillic acid. Oxidation of (71) with CrO_3-py afforded monoketone (72) with spectral data (λ_{max} 248 nm; 1700, 2480, 3200 and 2600 cm^{-1}) characteristic of a hydroxyl and an α,β-unsaturated cyclohexanone, indicating that (71) had one secondary and one tertiary hydroxy group. The position of the latter at C-7 can be ascertained from the multiplicity and location of the signal for the methyl group at 1.19 ppm in the ^{1}H NMR spectrum of (71).

Bicyclic alcohol esters with carotane skeletons are, as we have said, the most abundant sesquiterpenes in the genus *Ferula*. In order to give a better-organised account of the chemistry of these sesquiterpenes we have divided them for the purposes of this review into two groups: bicyclic alcohol esters hydroxylated at C-10 and those hydroxylated at C-6. The most important endogenous characteristics of the genus *Ferula* are the number of sesquiterpenes and coumarins to be found therein.

1. Bicyclic Alcohol Esters with a Hydroxy Group at C-10

In 1973 SUKH DEV's group (43) isolated a new sesquiterpene diol called jaeschkeanadiol from the roots of *Ferula jaeschkeana* Vatke and established its absolute stereostructure as (73).

Jaeschkeanadiol was the major component of both the light petroleum and acetone extracts and had a molecular formula of $C_{15}H_{26}O_2$ [M-H$_2$O]$^+$ at m/z 220. Its IR spectrum showed strong OH absorption. The D-exchanged ^{1}H NMR spectrum showed one secondary and one

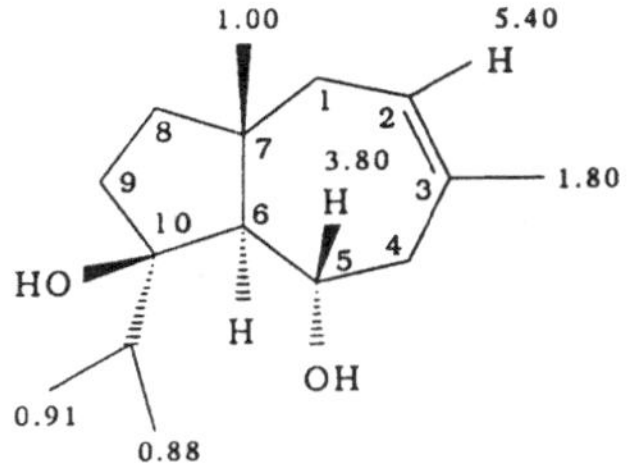

^{1}H NMR (CCl$_4$) and numbering system of (73) (*43*)

tertiary hydroxyl. When the compound was acetylated, a monoacetate was obtained with OH absorption still discernible in its IR spectrum.

Se-dehydrogenation of jaeschkeanadiol (**73**) gave aromatic products in considerable quantities from which preparative TLC separated dauca-lene (**74**), a typical dehydrogenation product for daucalene (**75**)-skeleton sesquiterpenes, such as, for instance carotol (**76**), (*44, 45*). Thus, it can reasonably be supposed that jaeschkeanadiol has the bicyclic framework (**75**).

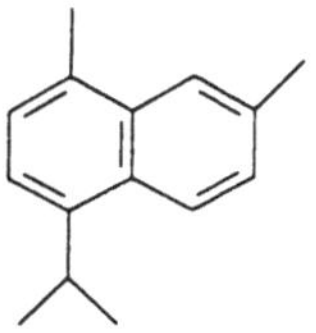

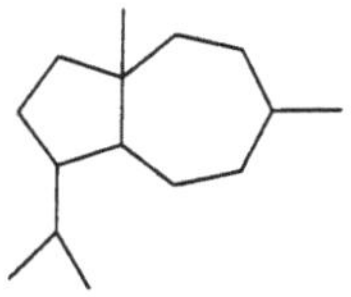

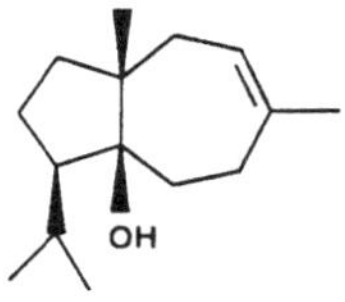

Bearing in mind the nature of the double bond of the diol and the presence of an isopropyl group in the molecule, there are two possible positions for the tert-OH (viz. C-6 and C-10) on the framework (**75**) for jaeschkeanadiol. Mass spectrum of the diol showed its base peak at m/z 195 [M-43]$^+$ suggesting loss of an isopropyl group. The easy elimination of this group (*46*) is best accounted for by the tertiary OH on C-10. The secondary hydroxy group can be sited on C-5 if dihydrojaeschkeanadiol is obtained and oxidized with Jones' reagent (*47*) to give a hydroxyketone which, on exposure to base (5% ethanolic KOH), will suffer water loss, affording an α, β-unsaturated ketone. This transformation clearly estab-lishes that hydroxyls are situated 1,3 to one other. The ^{1}H NMR spectrum of the α, β-unsaturated ketone does not show any absorption for an olefinic proton, and so must be (**77**). These considerations indicated that jaeschkeanadiol had one of two gross structures, (78) or (79). However, the appearance of the signal of the proton geminal to the secondary hydroxyl in the ^{1}H-NMR spectrum of the diol was conclusive. The multiplicity of the signal [triplet of doublets centred on $\delta 3.87$ $(J_1 = 10.0$ Hz, $J_2 = 5.0$ Hz)] was consistent only with structure (79).

(77)   (78)   (79)

The stereochemistry of jaeschkeanadiol (73) was established by a direct chemical correlation with laserol (80) (43, 48) via a degradation product common to both (Chart 9).

Oxidation of (83) with RuO_4 directly and surprisingly furnished the target compound (81) after the usual work-up involving extraction with 5% aqueous Na_2CO_3, instead of the expected carbonate (85). Conceivably β elimination is triggered by the CO group at C-3 to give (81) directly with loss of CO_2 (49).

Keto acid (81) proved identical with one obtained earlier by SORM et al. (48) from laserol (80).

As can be seen, the correlation between (73) and (80) clearly confirmed the stereochemistry of all asymmetric centres in jaeschkeanadiol (84) except for that at C-5. The configuration at C-5 was deduced by studying the carbonate (82) using Dreiding molecular models for the two possible configurations and showed that with C-5 either α-OH or β-OH, the cyclic carbonate ring could readily be constructed and was free of angle strain. However, presence of this ring placed further restraints on the number of energetically preferred conformations; in fact, given the configuration at C-5, there appeared to be only one preferred conformation. The seven-membered ring has a chair-like conformation in which the bonds at C-1, C-2, C-7, C-6 and C-5 have quasi axial-equatorial dispositions. The signal in the ^{1}H NMR spectrum of the carbonate appeared as a triplet of doublets with $J_1 = J_3 = 10.5$ Hz and $J_2 = 5.5$ Hz. This pattern of lines is consistent only with A, which has two axial-axial and one axial-equatorial type couplings (50). Therefore, jaeschkeanadiol was assigned stereostructure $C \equiv (73)$ which agrees with its most probable conformation as well.

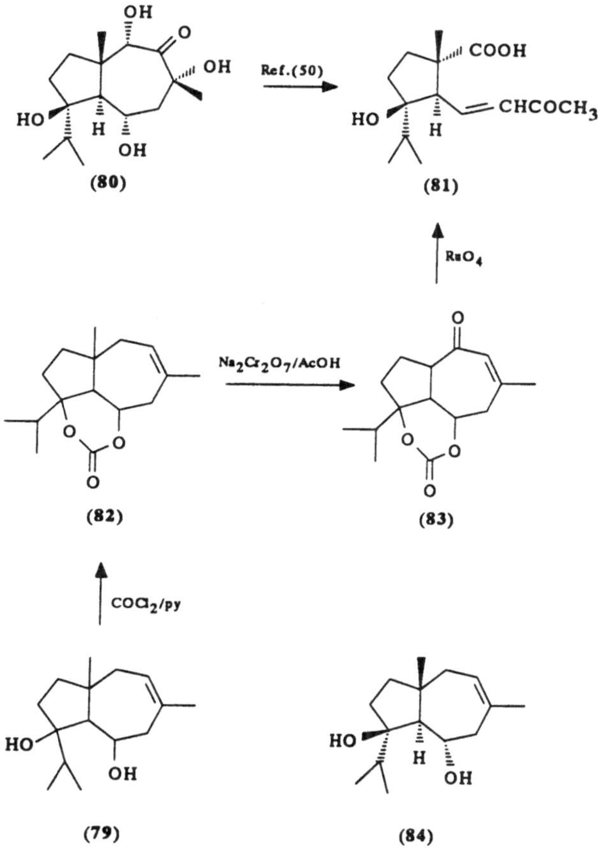

Chart 9. Correlation of jaeschkeanadiol and laserol

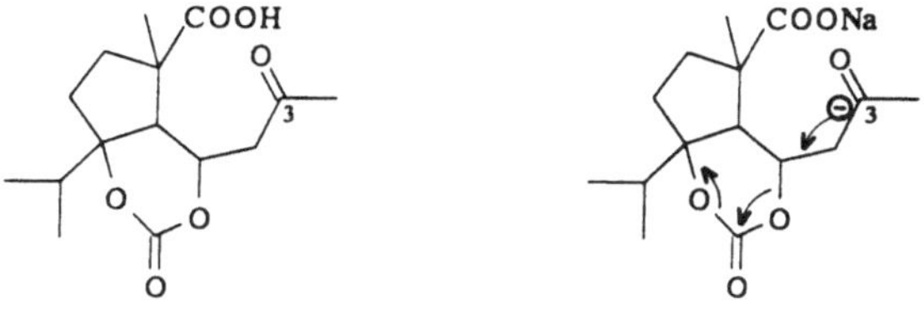

(85) (86)

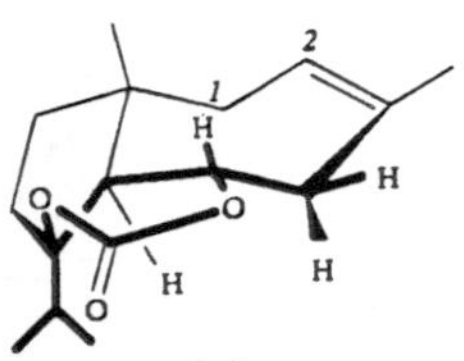

(a)

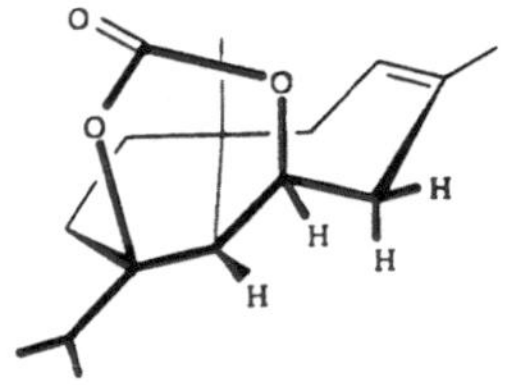

(b)

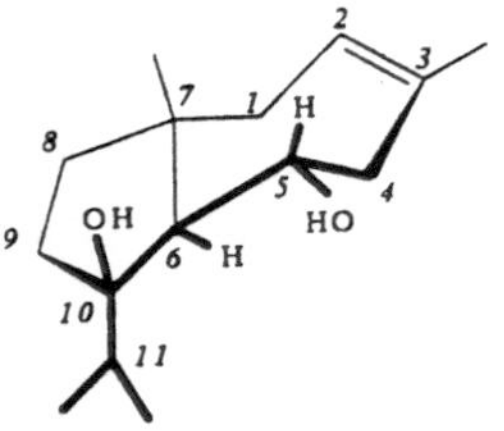

(c)

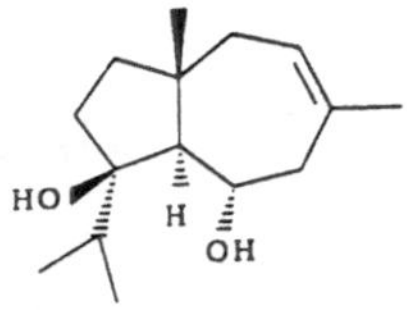

(73)

From the same species, S. N. Garg *et al.* (*51–53*) obtained seven new carotane sesquiterpenes, (**115**) (**117**), (**121**), (**128**), (**129**), (**131**), (**132**) and one new isocarotane sesquiterpene (**154**), all identified from their chemical and spectral data.

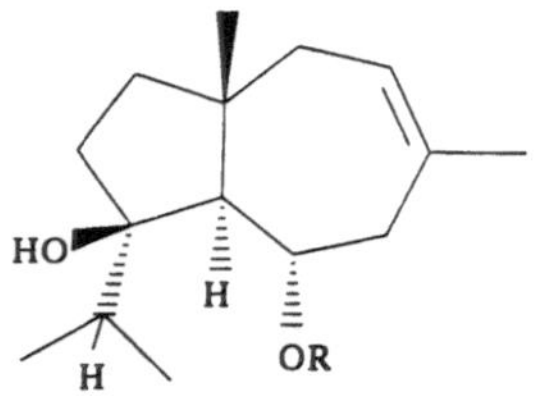

(**87**)   R= *p*-HyBz

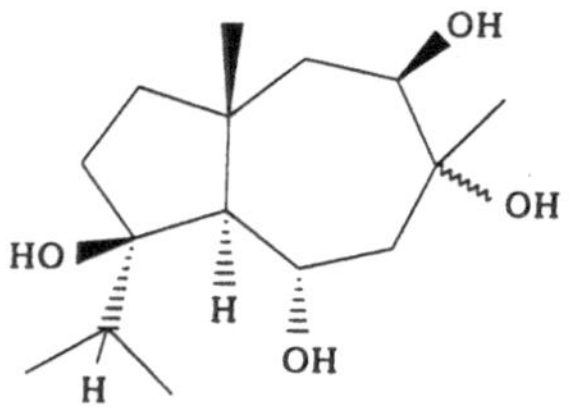

(**92**)

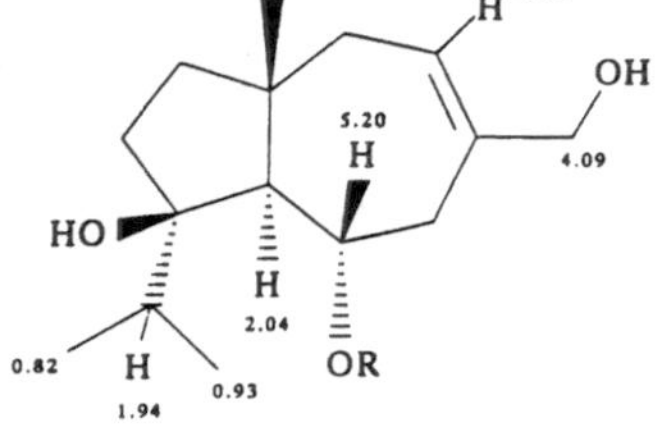

(**115**)   R= *p*-HyBz

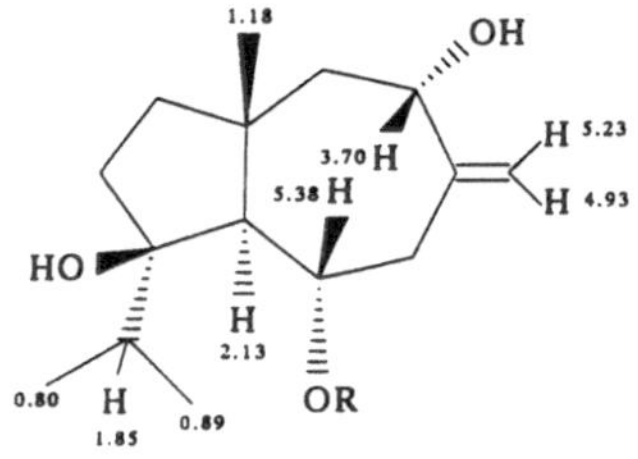

(**117**)   R= *p*-HyBz

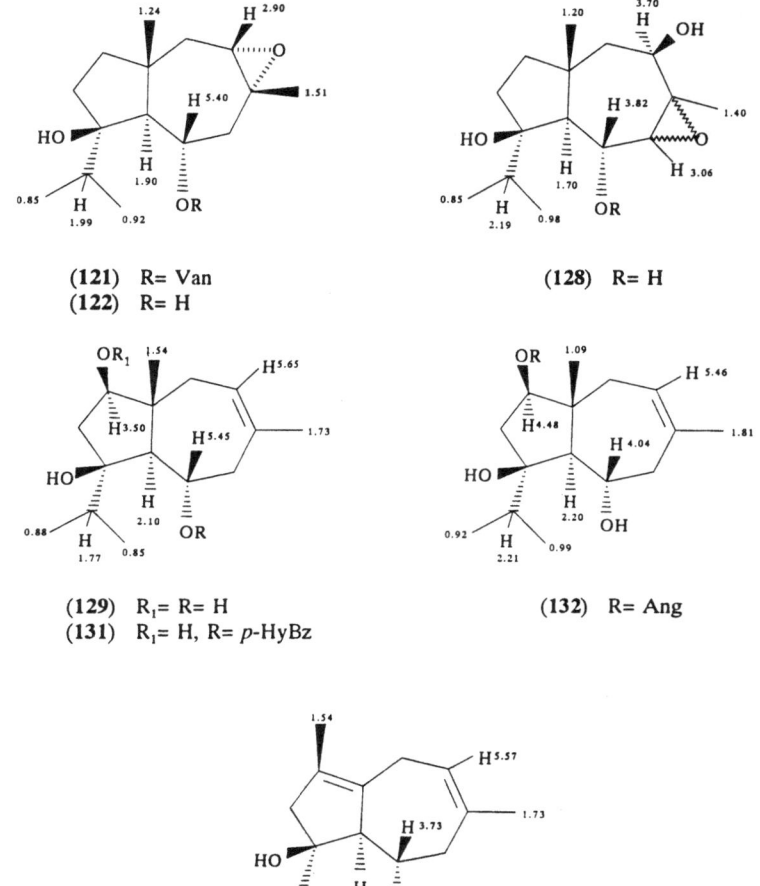

(121) R= Van
(122) R= H

(128) R= H

(129) R₁= R= H
(131) R₁= H, R= *p*-HyBz

(132) R= Ang

(154)

Feruginidin **(115)** and ferugin **(117)** have UV maxima at 256 and 258 nm, respectively, IR bands at 3300–3350 (OH) and 1680 cm^{-1} (ester) and two pairs of *ortho* coupled aromatic protons showing them to be aromatic esters. *p*-Hydroxybenzoic acid was obtained from both compounds as an alkaline hydrolysis product. A fragment of m/z 331 [M-C₃H₇]$^+$ in the mass spectra of **(115)** and **(117)** reflected the loss of an isopropyl group, a typical phenomenon in carotane sesquiterpenes with a hydroxy group at C-10 (46). The ^{1}H NMR spectra of (115) and (117) were analogous to that of ferutinin (87), isolated earlier from *F. elaeochytris* (54). Compound (115) had a signal for a hydroxy methyl group at δ4.09 and not the vinyl methyl group at δ1.84 seen in (87). A secondary hydroxy group at C-2 in (117) was assigned α stereochemistry on the basis of coupling constants of 6.0 and 9.0 Hz between H-1β and H-2β, and H-1α, respectively. Signals were also observed in this compound for an exo-methylene group at δ4.93 and 5.23 instead of the vinyl methyl group possessed by (87).

Substance (121) formed a monoacetate and, when subjected to alkaline hydrolysis, yielded an acid identified as 4-hydroxy-3-methoxy-benzoic acid and an alcohol already obtained previously as a natural product from *F. linkii-GC* (55) and identified as 2,3-epoxyjaeschkeanadiol (122). The mass spectrum of (128) had a peak at m/z [M-C$_3$H$_7$]$^+$ but no molecular ion. Its ^{1}H NMR spectrum revealed the presence of two secondary hydroxyls, one epoxide and a methyl geminal to an oxygen. Positions of one of the secondary hydroxy groups and the epoxide ring were determined on the basis of chemical shifts (56) and a series of ^{1}H NMR homo-decouplings at C-5 and between C-3 and C-4. The hydroxyl was sited at C-2 since no acetonide was formed when (128) was treated with Me$_2$CO and anhydrous CuSO$_4$, and it also failed to react with sodium periodate, ruling out C-8. The C-9 position was also eliminated because the signal of the oxymethine proton appeared at δ3.70. Reduction of (128) with LiAlH$_4$ yielded (92) which consumed sodium periodate and confirmed that the hydroxy group should be at C-2, and the relative stereochemistry of the oxygen groups at C-2 and C-5 was ratified as β and α, respectively (57). The stereochemistry of the oxirane ring, however, has not yet been determined.

Preliminary studies of compounds (129), (131) and (132) (IR, ^{1}H NMR and MS) indicated that they were carotane derivatives. The mass spectra of these three compounds all had a fragment at [M-C$_3$H$_7$]$^+$ produced by the loss of an isopropyl group. Substance (129), akichenol, was reported for the first time as a natural product, having previously been obtained as a hydrolysis product of jaeschferin (58). Compounds (131) and (132) yielded akichenol (129) and *p*-hydroxybenzoic acid, and akichenol and angelic acid, respectively, when hydrolyzed with alcoholic potassium hydroxide. The upfield shift of the C-5 oxymethine signal in the ^{1}H NMR spectrum of (131) (from δ5.49 to 3.93) showed that this acid esterified the C-5 hydroxyl of akichenol while the upfield shift of the C-9 oxymethine signal from δ4.48 to 3.30 demonstrated that the acid in (132) was attached to the C-8 hydroxyl of akichenol.

Ferujaesenol (**154**) on the basis of its ^{1}H and ^{13}C NMR spectra had two tri- and tetrasubstituted double bonds and two hydroxy groups, one secondary and the other tertiary. If (**154**) were assumed to have a carotane skeleton, the positions of the trisubstituted double bond between C-2 and C-3 and the secondary hydroxy group at C-5 could be fixed by a series of decoupling experiments. Signals for another olefinic methyl in the ^{1}H and ^{13}C NMR spectra of (**154**) at δ 1.54 and 25.73, respectively, indicated that it was an isocarotane derivative in which the methyl had migrated to C-8 and that there was a double bond between C-8 and C-7. The alternative placement of this double bond, between C-1 and C-7, was discounted because the UV spectrum did not reveal the corresponding conjugation. The stereochemistry at C-5 and C-6 was assigned in the same way as for jaeschkeanadiol (**73**) and akichenol (**129**).

The new sesquiterpene ketoalcohol feruone (**159**) (*57*) showed absorption bands for a hydroxy group (3340 cm^{-1}) and an α,β-unsaturated ketone (1660 cm^{-1}) which was also indicated by the UV absorption at 238 nm. The ^{1}H NMR spectrum showed signals for a secondary oxymethine, an isopropyl group, an angular methyl and an olefinic methyl attached to a trisubstituted double bond conjugated with the keto group. The α configuration was assigned to the hydroxy group at C-2 on the basis of the H-2 coupling constant ($J_{1,2} = 10.0$ Hz) which would have been greater if the hydroxyl had been β.

(**159**)   (**160**)   R = *p*-HyBz

The IR spectrum of (**160**) had absorptions for a hydroxyl (3380 cm^{-1}) and an ester carbonyl (1690 cm^{-1}), while UV absorption at 258 nm and a MS fragment at 218 [M $-$ H$_2$O–C$_6$H$_4$(OH)COOH]$^+$ pointed to its being a *p*-hydroxybenzoic acid ester. When it was saponified with alcoholic KOH, *p*-hydroxybenzoic acid and an alcohol, ferutriol, (*56*) were obtained. The C-3 stereochemistry was described as shown in the formula without supporting evidence for the assignment (*59*).

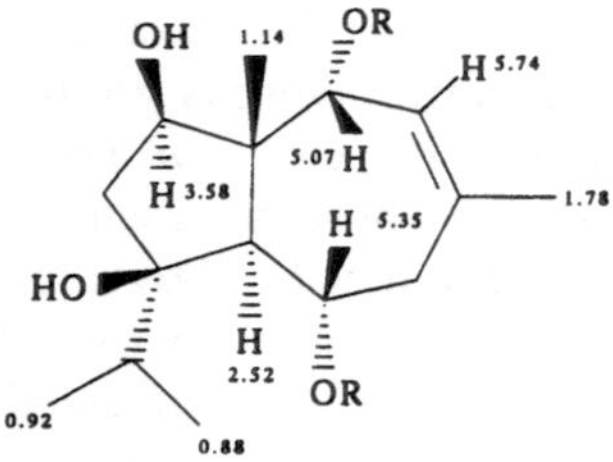

(161)   R = Ang

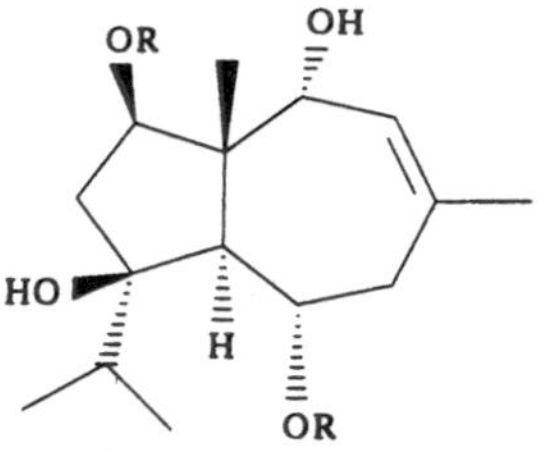

(162)   R = Ang

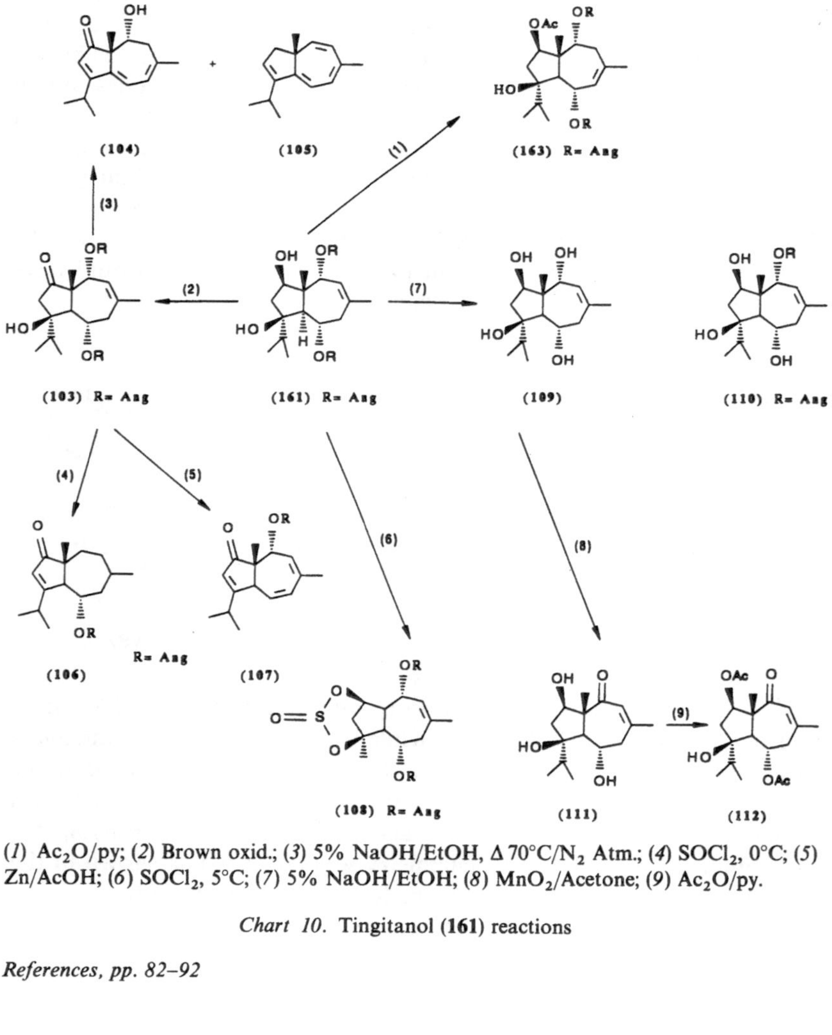

(1) Ac₂O/py; (2) Brown oxid.; (3) 5% NaOH/EtOH, Δ 70°C/N₂ Atm.; (4) SOCl₂, 0°C; (5) Zn/AcOH; (6) SOCl₂, 5°C; (7) 5% NaOH/EtOH; (8) MnO₂/Acetone; (9) Ac₂O/py.

Chart 10. Tingitanol (**161**) reactions

References, pp. 82–92

From *F. tingitana* MISKI *et al. (60)* isolated a new sesquiterpene ester, tingitanol (**161**), which at first they assumed to be identical with a known compound, deoxodehydrolaserpitine (**162**) obtained earlier from *Laserpitium latifolium* L. by HOLUB *et al. (61)*. However, comparison of (**161**) with an authentic sample of (**162**) showed that one of the two angeloyloxy moieties was at C-3 and not at C-4 as in (**162**). The series of chemical reactions carried out to determine the skeleton and the relative positions of the hydroxy groups is shown in Chart 10.

X-ray analysis of acetyltingitanol (**163**) confirmed the suggested structure and the relative stereochemistry of compound (**161**) *(60)*. Later *(62)*, (**163**) was isolated from the same *Ferula* species together with acetyldeoxodehydrolaserpitine (**164**) and 10β-hydroxy-5α-*p*-hydroxybenzoyloxy-1α-angeloyloxy-dauc-2-ene (**165**). The ¹H NMR spectrum of (**164**) was similar to that of (**163**) while that of (**165**) resembled the one recorded for (**166**) from *Ferula communis (63)*.

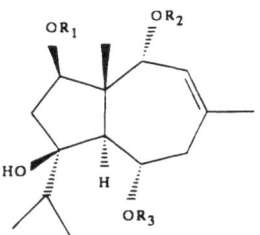

(**163**) $R_1 = Ac, R_2 = R_3 = Ang$	(**165**) $R_1 = Ang, R_2 = p\text{-HyBz}$
(**164**) $R_1 = R_3 = Ang, R_2 = Ac$	(**166**) $R_1 = Ang, R_2 = p\text{-Anis}$

In this context it is noteworthy that *F. tingitana, F. communis, F. linkii* and *F. lancerottensis*, all members of subgenus Euferula (Boiss.) Korovin, yielded both 1,5 *cis*- and *trans*-daucane derivatives *(64–67)* which can be biogenetically related as shown in Chart 11.

These biogenetic considerations require the oxirane ring of (**116**) to be β-oriented, on the basis of the following correlation between the ¹H NMR spectra of (**116**), jaeschkeanadiol (**73**) and carotol (**76**). The nearly identical chemical shifts of the isopropyl methyl groups of (**116**) (δ 17.5 and 18.5) and (**73**) (δ 17.8 and 18.2) in contrast to those of carotol (**76**) (δ 20.9 and 23.5) indicated that the sesquiterpene nucleus and the C-4 asymmetric centre exercised similar shielding and γ-substituent effects, respectively, on this part of the molecule; these, in turn, suggest a β-stereochemistry for the epoxy group in (**116**).

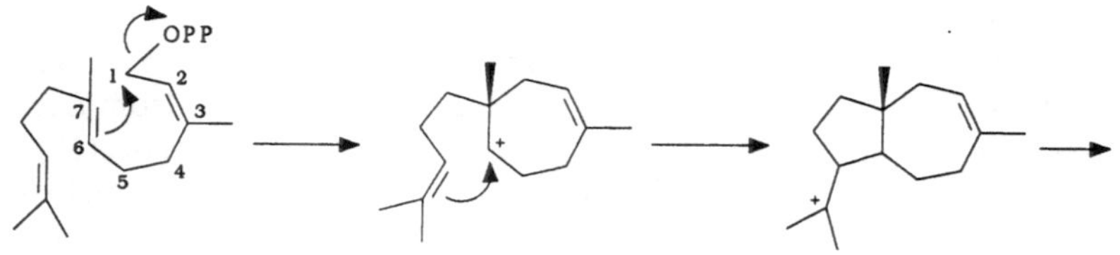

Farnesyl pyrophosphate

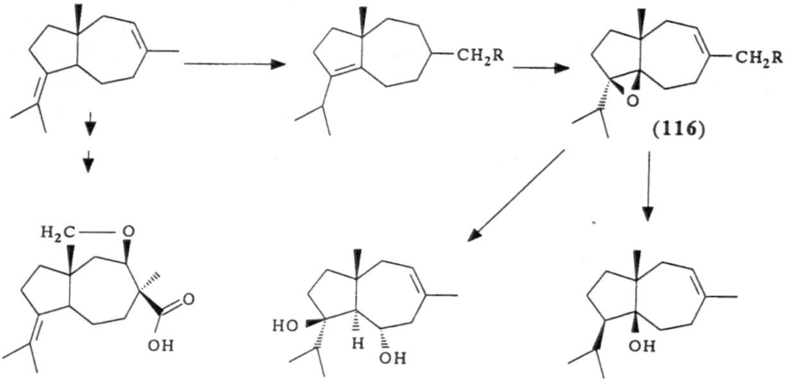

(116)

Aspterric acid

(73) Jaeschkeanadiol
(1,5-*trans*-daucane)

(76) Carotol
(1,5-*cis*-daucane)

Chart 11. Biogenetic relationship of 1,5-*cis* and 1,5-*trans*-daucanes in *F. tingitana*, *communis*, *linkii* and *lancerottensis*

The new sesquiterpenes **(138)**, **(180)**, **(185–190)**, **(199)** and **(209)** were isolated *(68, 69)* from *Ferula sinaica* and their structures were determined by study of their ^{1}H and ^{3}C NMR spectra. The ^{1}H NMR data of **(180)** appeared to be consistent with those of **140**, a compound isolated from *F. communis (63)*, with one significant difference—two doublets at δ 0.84 and 0.90, indicative of an isopropyl moiety. It is known *(63)* that the doublets representing the isopropyl methyls are farther apart (ca 0.13 ppm) in compounds with an aromatic acyl group at C-5α, such as **(140)**. The 0.06 ppm separation in **(180)** therefore indicated that the aromatic acyl group was to be found at C-8α in this substance.

The ^{1}H and ^{13}C NMR spectra of isofercomin **(185)** were very similar to the corresponding spectra of fercomin from *F. communis (64)*. The main difference between the two compounds was the position of the hydroxy

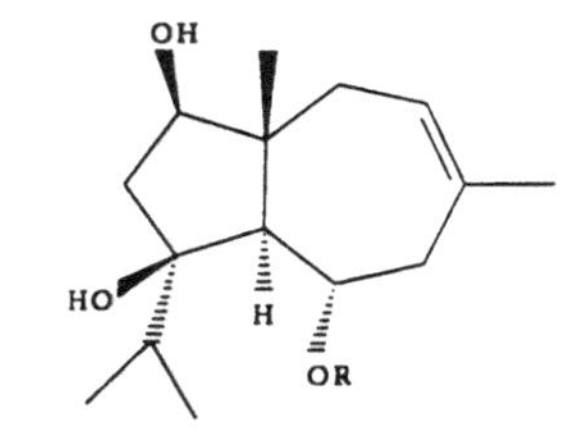

(138) R = *p*-Anis

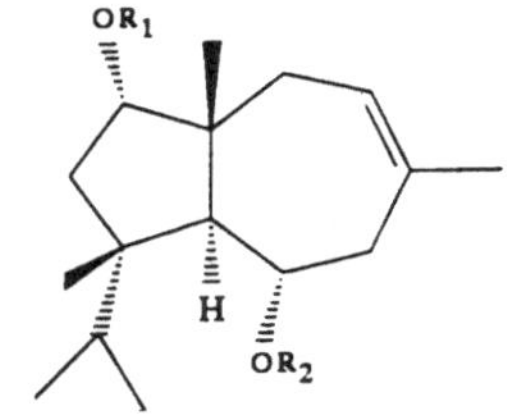

(140) R₁ = Ac, R₂ = Bz
(180) R₁ = Bz, R₂ = Ac

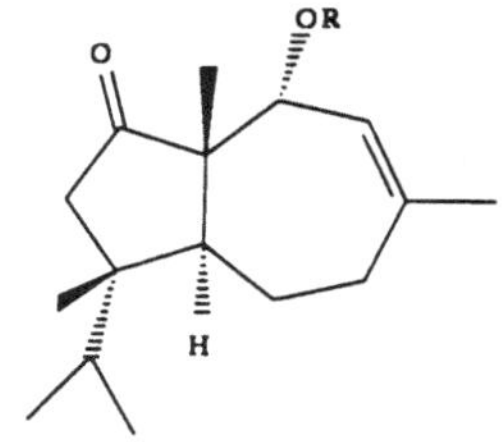

(185) R = *p*-Anis

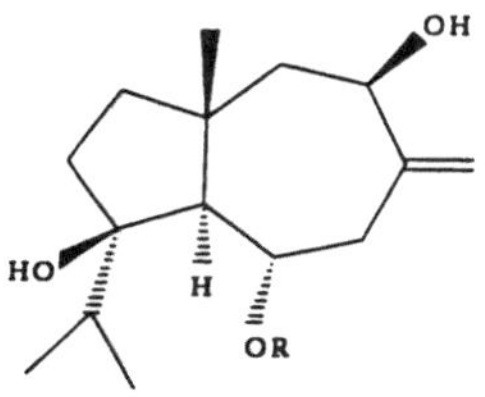

(186) R = *p*-HyBz

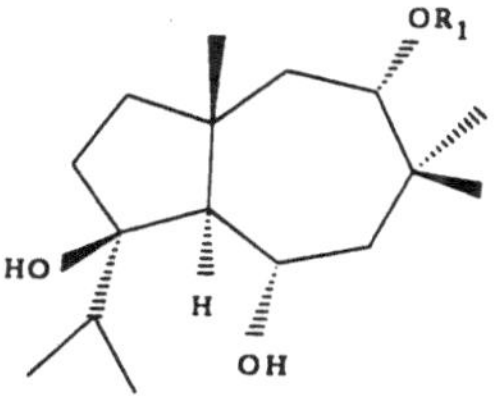

(187) R = *p*-HyBz

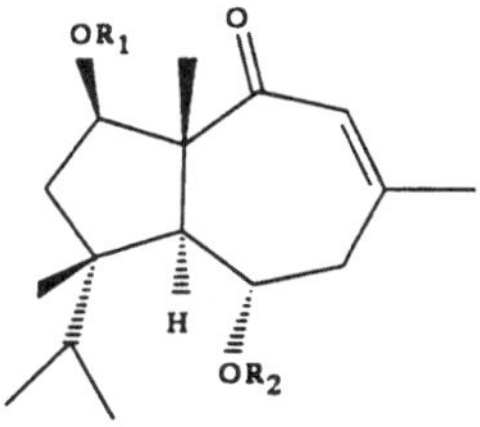

(189) R₁ = R₂ = H
(190) R₁ = *p*-HyBz, R₂ = H
(199) R₁ = R₂ = Ang

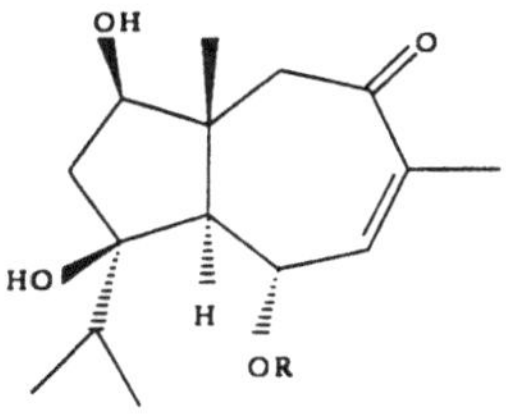

(209) R = *p*-HyBz

group, not at C-6 as in fercomin (*64*) which would involve a ^{13}C NMR shift of δ 60.21, but at C-10 (δ 86.40).

The structure of 2-*epi*-5-*p*-hydroxybenzoylisolancerotriol (**186**) was ascertained by comparing its ^{1}H NMR spectral data with those for other isolancerotriol esters (*67, 70*). The two doublets representing two protons each at δ 7.95 and 6.88 indicated that a *p*-hydroxybenzoyl group was present, and a broad signal at δ 4.50 for the H-2 confirmed the β-orientation of the hydroxy group. The ^{1}H NMR spectrum of (**187**) was very similar to that of isomer (**188**), obtained earlier from the same species (*59*). Other compounds identified were 5α,10β-dihydroxydauc-2-ene-1-one (**189**), 5α-*p*-hydroxybenzoyloxy-10β-hydroxydauc-2-ene-1-one (**190**) and 5α,8β-diangeloyloxy-10β-hydroxydauc-2-ene-1-one (**199**), all with similar ^{1}H NMR data. Four aromatic protons appeared as two doublets at δ 7.92 and 6.90 in (190), a characteristic of a *p*-hydroxybenzoyl group. The olefinic proton recorded as a broad singlet at δ 6.03 showed only long-range coupling with the olefinic methyl at δ 2.02. The downfield shift of both signals indicated a keto group in the neighbourhood (IR, 1650 cm^{-1}). Methanolysis of (190) gave (189).

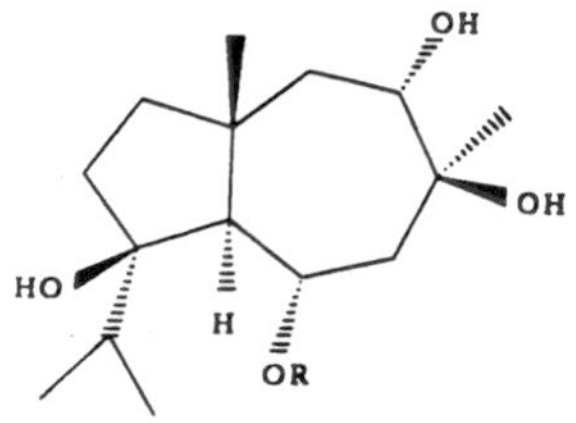

(188)

Compound (199) had two angelates at C-5 and C-8. The stereo-chemistry of the asymmetric centres of (189), (190) and (199) was defined by comparing chemical shifts and coupling constants with those of tingitanol (100), isolated, as stated above, from *F. tingitana* (60). The structure of (209) was deduced from its ^{1}H NMR spectrum which was very close to that of the known substance 213 (67).

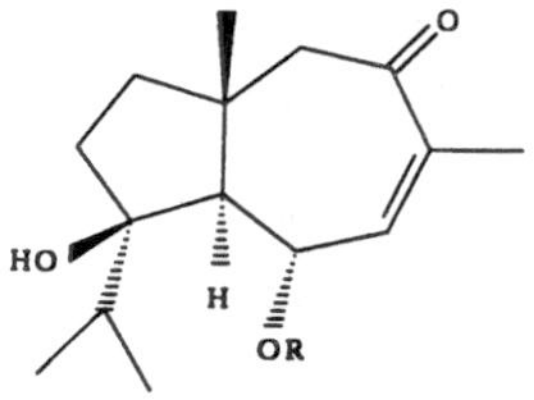

(213)

The mass spectrum of (138) showed loss of water, an isopropyl radical and anisic acid at *m/z* 370, 345 and 236, respectively. Its structure was determined using the spin decoupling technique and its stereochemistry confirmed by comparing the chemical shifts and coupling constants with compounds reported from *F. communis* var. *brevifolia* and *F. jaeschkeana* (71, 72).

2. Bicyclic Alcohol Esters with a Hydroxy Group at C-6

The carotanes (daucanes) belonging to this small group of sesquiterpenes with a hydroxyl at C-6 are derived from one simple compound,
carotol (**76**) (*73*).

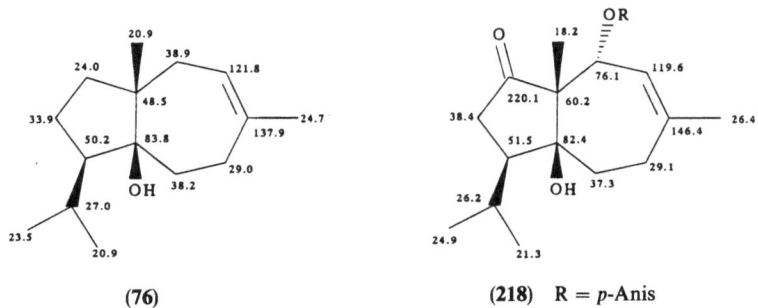

(76)   **(218)**   R = *p*-Anis

Chart 12. ^{13}C NMR data (*74*)

As mentioned earlier in the section on isofercomin, MISKI *et al.* (*64*)
isolated a new carotane ester fercomin (**218**) from *F. communis* subsp.
communis as well as a new lactone, fercolide (**220**). Its IR spectrum
indicated the presence of hydroxyls (3250, 1030 cm^{-1}), a cyclopentanone
(1740, sh, 1730 cm^{-1}) and an aromatic ester (1710, 1608, 1580, 1510 and
1260 cm^{-1}). Its MS and ^{13}C NMR data defined (**218**) as a bicyclic
structure with one five- and one seven-membered ring. The ^{1}H NMR
spectrum and spin decoupling experiments showed a tertiary methyl
signal at δ 1.10, two isopropyl methyl doublets at δ 1.02 and 1.14 and a
doublet characteristic of a proton under an ester function at δ 5.72, which
was coupled only with a broad vinyl proton doublet at δ 5.46. The latter
in turn was coupled allylically with a vinyl methyl at δ 1.76. These details
identified the structure of fercomin (**218**) as 8-keto-6-hydroxy-1-*p*-
anisoloxy-dauc-2-ene. Its stereochemistry was assigned by comparing its
^{13}C NMR data with those of carotol (**76**), a compound with a stereochemistry soundly established by single-crystal X-ray analysis (*74*) and
total synthesis (*75, 76*) (Chart 12). The two compounds were seen to have
the same stereochemistry at C-7, C-10 and C-6; this was confirmed by an
X-ray analysis. When the ^{13}C-NMR spectra of (**218**) and related compounds were compared with those reported for vaginatin, a substance
reported earlier from *Selinum virginatum* (Apiaceae) (*151*) and *Inula
crithmoides* (*77*), the structure of the latter was revised to (**130**) (*64*).

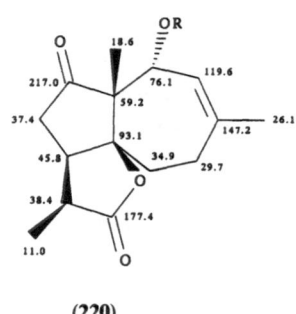

(130)

Fercolide **(220)**, an amorphous solid with a saturated γ-lactone (IR band at 1775 cm^{-1}) and a ketone on a five-membered ring (1746 cm^{-1}), had the molecular formula $C_{23}H_{26}O_6$ and contained the same two acyl groups as **(218)**, but one of the secondary methyls on the isopropyl was oxidized to a carbonyl and apparently engaged in lactone ring formation with the hydroxyl on C-6.

The presence of a lactone carbonyl signal (177.4 ppm) in the ^{13}C NMR spectrum of **(220)** (Chart 13) together with changed chemical shifts for C-6, C-10, C-11 and C-12 of the molecule in comparison to those of fercomin **(218)** confirmed the lactone system; the ^{1}H and ^{13}C NMR spectra clearly established that the methyl group on C-11 was β-oriented.

(220)

Chart 13. ^{13}C NMR spectrum of fercolide **(220)** *(64)*

This is perhaps the place to mention that several other common types of sesquiterpene lactones (e.g., germacranolides, eudesmanolides, guaianolides and eremophilanolides) are frequently encountered in the *Apiaceae* *(78)*. Those isolated from *Ferula* species are listed in Section VIII, Table 2 and Chart 19, but no attempt will be made to discuss their structure elucidation. However although daucane-type sesquiterpenes are obtained in the greatest abundance from the *Apiaceae* (*48, 79*), lactone (**220**) remains the only daucane lactone isolated so far from this family.

In the opening paragraphs of this review we referred to three well-characterized *Ferula* species found in the Canary Islands: *Ferula linkii* Webb, *Ferula latipinna* Santos and *Ferula lancerottensis* Parl. *Ferula linkii* is the most widely distributed of these, not only throughout the archipelago but also at island level, and occupies quite a range of ecological habitats. It can be found anywhere from middling altitudes (200–300 m) to over 2000 m on the mountains of the Teide National Park in Tenerife. When the phytochemistry of one population of *F. linkii* collected in the Barranco de Guayadeque (Southern Gran Canaria) (*55*) was compared with the chemistry of another population from North Tenerife (*80*) significant differences were observed, to such an extent that there would seem to be a closer phytochemical relationship between *F. linkii* from Tenerife and *F. latipinna* from La Palma (*81*) than between the two species of the same name.

Detailed examination of the taxonomic characters of the Tenerife collection from the slopes above El Guincho, between Icod and Garachico, and comparison with those growing in other northern locations in Tenerife, from the Anaga massif to Teno, yielded no significant differences which would distinguish it from *F. linkii* and establish it as a new taxon. Unfortunately, no voucher specimen of the Gran Canary collection is on file (*82*). The question thus remains open and calls for further chemical and taxonomical investigation of this genus in the Islands. In view of the phytochemical results, it has been considered convenient to designate the two varieties as *F. linkii-TF* and *F. linkii-GC* to differentiate the populations from Tenerife and Gran Canaria.

The two other species, *F. latipinna* Santos and *F. lancerottensis* Parl., are endemic to the islands of La Palma and Lanzarote, respectively, with the former possibly occurring also in La Gomera (pending confirmation). Fig. 4 shows the distribution of these *Ferula* species in the Canary Islands.

In the first publication on *Ferula linkii-GC*, GONZALEZ et al. (*83*) reported a new sesquiterpene ester, linkiol (**221**), hydroxylated at C-6. Linkiol (**221**), $C_{20}H_{34}O_4$, had characteristic alcohol and ester group bands in the IR spectrum; its structure and stereochemistry were ascertained from the fact that the sequence (**221**) → (**133**) → (**134**) → (**135**) → (**136**) → (**137**) could be carried out.

In a daucane structure, a tertiary alcohol group will be found at C-3, C-6 or C-10 and a secondary one at C-2 or C-4. In the ^{1}H NMR spectrum of linkiol, the resonance of the isopropyl methyls was not in the same

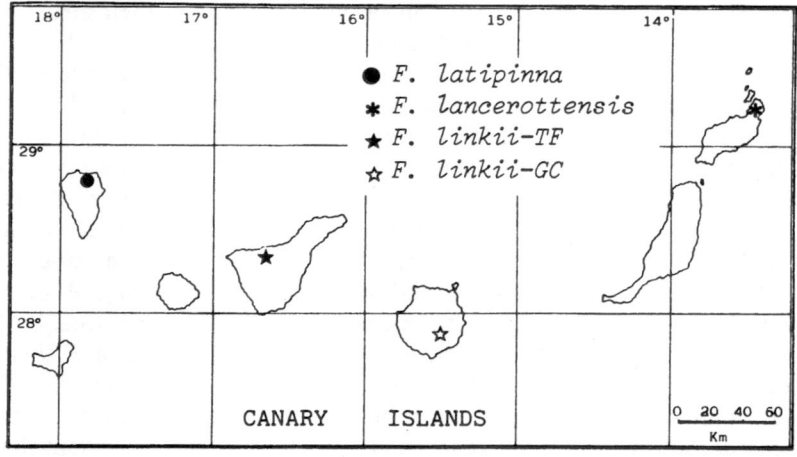

Fig. 4. Approximate distribution of the *Ferula* taxa studied to date

position (δ 0.95, *t*, 6H) as in jaeschkeanadiol (**73**) which has a hydroxyl at
C-10. The position at C-6 could therefore be assigned to a second tertiary
hydroxyl while formation of an acetonide (**134**) indicated that a secondary
hydroxyl was located at C-2 or C-4. In order to decide which of these two

Reagents: (*i*) 5% KOH in methanol; (*ii*) dry Me$_2$CO, CuSO$_4$; (*iii*) dry C$_5$H$_5$N, SOCl$_2$;
(*iv*) *p*-toluene-sulfonylchloride, dry C$_5$H$_5$N

References, pp. 82–92

positions was correct and to relate (221) to an already known compound, alcohol (133) on treatment with tosyl chloride in pyridine was converted to tosylate (136) and a compound identified as daucol (137) (*84*) produced by solvolysis of the tosylate in the reaction medium. When the tosylate was solvolysed with methanolic KOH, daucol (137) was obtained in pure form. Thus, the structure and stereochemistry of the carbon skeleton of sesquiterpene (133) could be established with both tertiary hydroxyls fully characterized and the secondary alcohol sited on C-2. The β-orientation of this secondary alcohol was determined from the chemical shifts seen in the ^{1}H NMR spectrum and the differences between alcohol (133) and triol (139), obtained from carotol (76) by treatment with potassium permanganate (*44*).

(76)   (139)

Two other sesquiterpenes isolated from *Ferula linkii-GC* were carotdiol acetate (223) and carotdiol veratrate (224) (*85*). Structures of these substances were determined on the basis of the following considerations. The high resolution MS of (223) was in accordance with formula $C_{17}H_{26}O_3$, while the IR spectrum showed absorbances characteristic of a hydroxyl (3590 cm^{-1}) and an ester (1720 cm^{-1}). The ^{1}H NMR spectrum indicated not only that this ester was an acetate and that the hydroxy group was of the tertiary type, but also exhibited signals typical of a vinyl proton, an isopropyl group and two methyls, one of the methyl groups being located on a double bond.

Carotdiol (142) was obtained by hydrolysis of (223), while oxidation of (142) with pyridinium dichromate afforded the α,β-unsaturated ketone (143). Carotdiol (142) could also be esterified with 3,4-dimethoxybenzoyl-chloride in pyridine to give a compound identical with natural carotdiol veratrate (224).

Structure (144) is an alternative to (142) but was ruled out when carotdiol veratrate (224) on oxidation with *m*-chloroperbenzoic acid gave (145) formed by the opening of the oxirane ring of the intermediate epoxide (146). This reaction sequence together with the ^{1}H NMR analysis

permitted location of the double bond of the two carotdiols (223) and
(224) on C-2, C-3 while formation of an ether (145) further confirmed the
placement of the tertiary hydroxy group at C-6. Epoxidation of (224) gave
a minor compound (147) which had a ^{1}H NMR spectrum similar to that
of (145) but without the geminal proton of the secondary hydroxy group.

   The stereochemistry of the ester group at C-4 in (223) or (224) (86) was
determined in the following manner. The Dreiding stereomodel of
carotdiol (142), assuming a conformation for the 7-membered ring which
would accommodate the coupling constants between the vinyl hydrogen
and the two protons at C-1 (J = 5.5 and 3.0 Hz), indicated that the proton
responsible for the broad singlet of the geminal proton of the alcohol

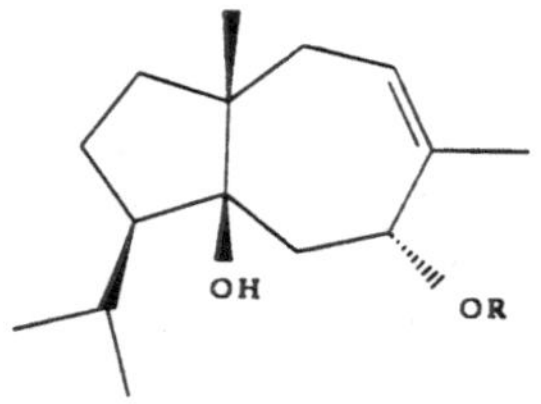

(142)   R = H
(223)   R = Ac
(224)   R = Ver

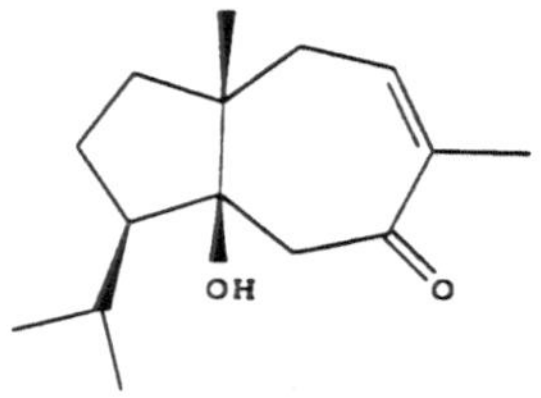

(143)

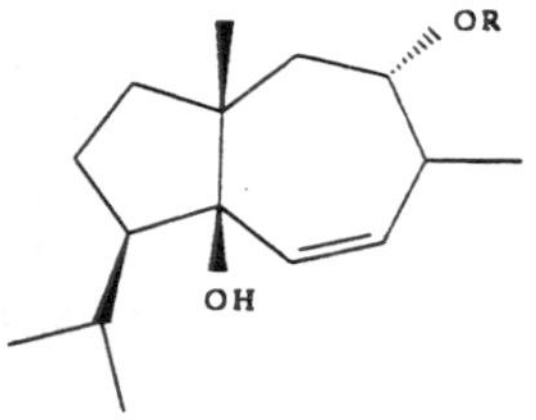

(144)

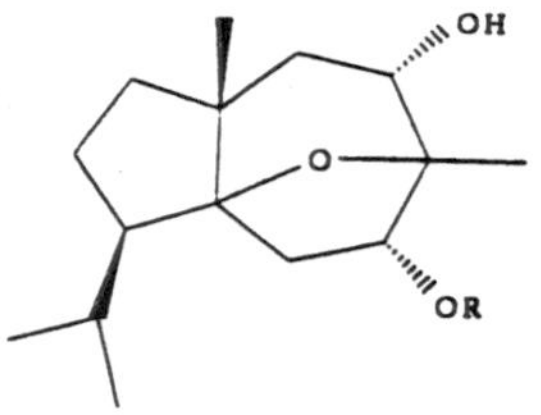

(145)   R = Ver

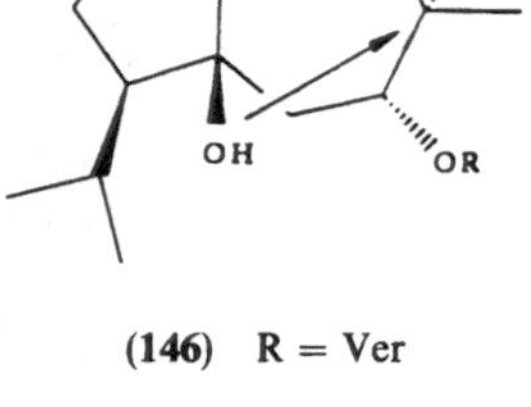

(146)   R = Ver

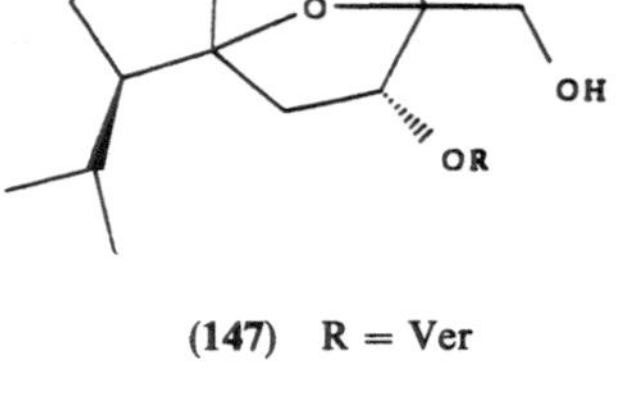

(147)   R = Ver

group at C-4 coupled with the two hydrogens at C-5 should be β-oriented and that, therefore, the hydroxyl at C-4 had to be α. This conformation for (**142**) or its esters was similar to that of carotol (**76**) and also explained formation of the daucol-type hydroxy ether (**145**) on epoxidation of (**224**) due to the spatial proximity of the hydroxy group at C-6 to the double bond. The alternative conformation would not permit formation of this type of ether (*86, 87*). Several other reasons were also adduced (*85*) for assigning α-stereochemistry to the alcohol group at C-4.

Three new daucanes (**123–125**) differing only in the acid esterifying a secondary hydroxy group were also isolated from *Ferula linkii-GC* (*55*). The NMR spectra exhibited signals typical of an isopropyl group, an angular methyl group and a methyl group attached to a carbon with an oxygen group which must form part of an oxirane ring since the signal of a proton geminal to the oxygen of this ring was also observed. One compound, epoxyjaeschkeanadiol *p*-methoxybenzoate (**123**) was identical with the epoxidation product of ferutidin (**88**) (*88*). As epoxidation of the seven-membered ring is known to take place from the α-face (*86*), the stereochemistry at C-2 and C-3 followed.

Another of the new esters was epoxyjaeschkeanadiol isovalerate (**124**) which, when hydrolyzed, gave an alcohol (**151**) and, after neutralization of the alkali excess dilute hydrochloric acid, an aldehyde, (**152**), as the result of acid-catalyzed opening of the epoxide ring and 1,3-cleavage of the resulting 1,3-diol.

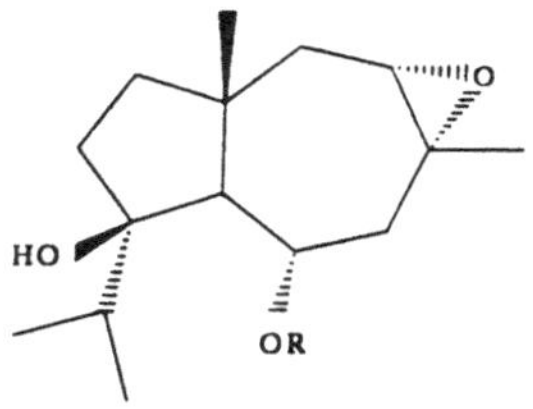

(**123**)   R = *p*-Anis
(**124**)   R = iVal
(**125**)   R = Ver
(**151**)   R = H

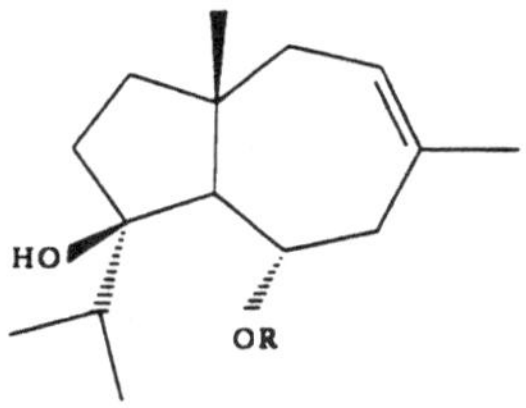

(**149**)   R = *p*-Anis

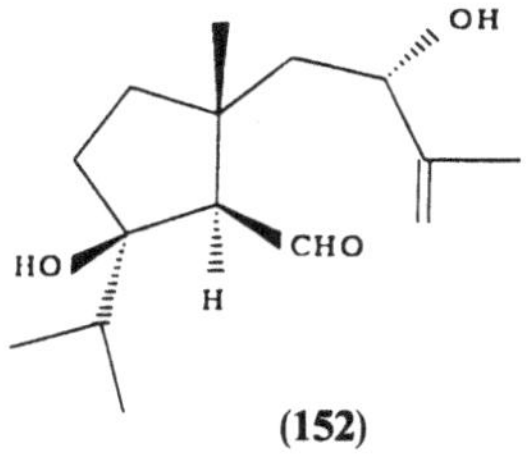

(**152**)

Other new constituents from *F. linkii-GC* (*56, 70*) include felikiol 3-angelate (**225**), webiol angelate (**227**) and epoxyangelate (**228**), ferutriol 5-isovalerate (**229**), lancerotol veratrate (**211**), isolancerotriol 5-isovalerate (**118**), isolancerotetrol 5-isovalerate (**230**) and 5-angelate (**231**) and epoxy-isolancerotetrol 5-isovalerate (**232**).

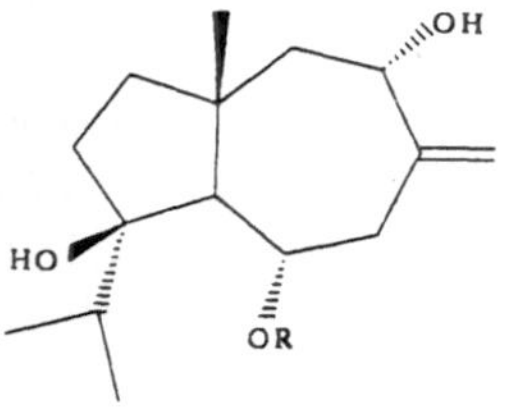

(**118**)   R = iVal

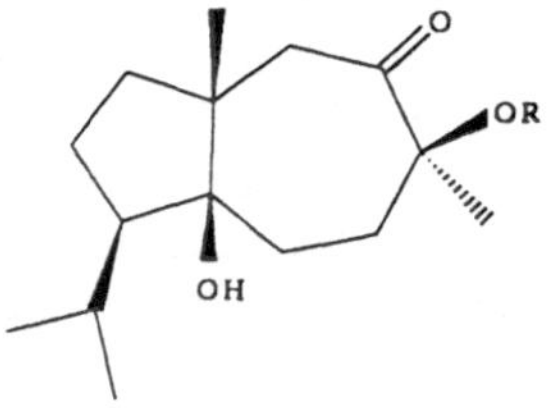

(**155**)   R = H
(**225**)   R = Ang

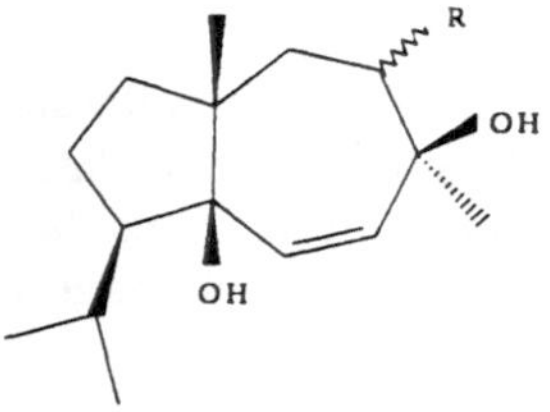

(**156**)   R = α-OH
(**157**)   R = β-OH

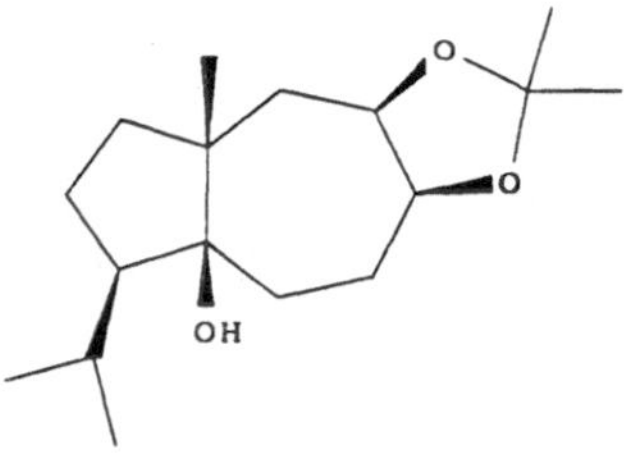

(**158**)

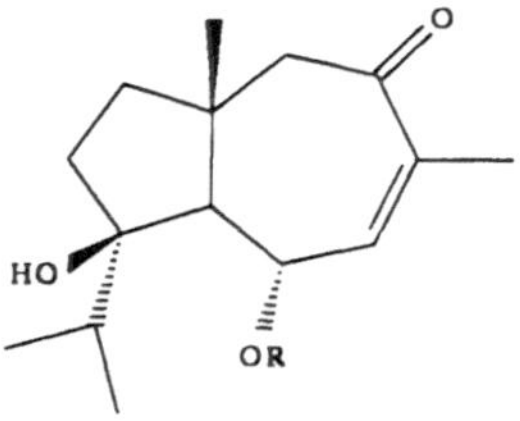

(**211**)   R = –COC₆H₃(OMe)₂

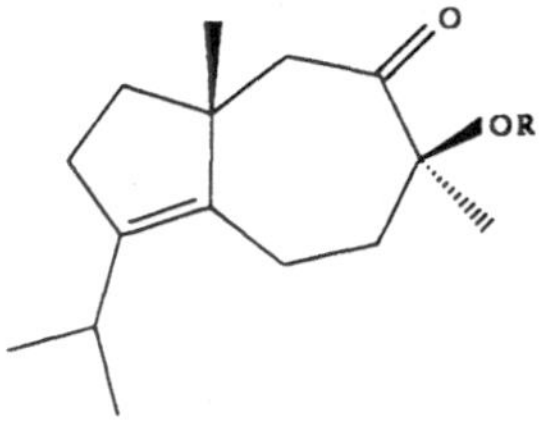

(**227**)   R = Ang
(**228**)   R = E-Ang

Structure (**225**) was assigned on the basis of the following consider-ations: the HRMS was in accordance with formula $C_{20}H_{32}O_4$ and the IR spectrum had bands for hydroxy, carbonyl and ester groups. The

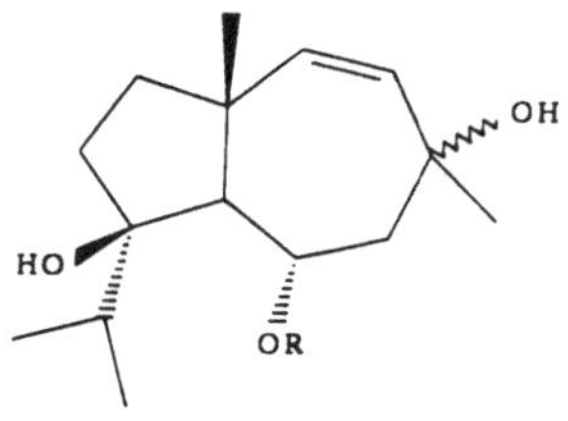

(229)   R = iVal

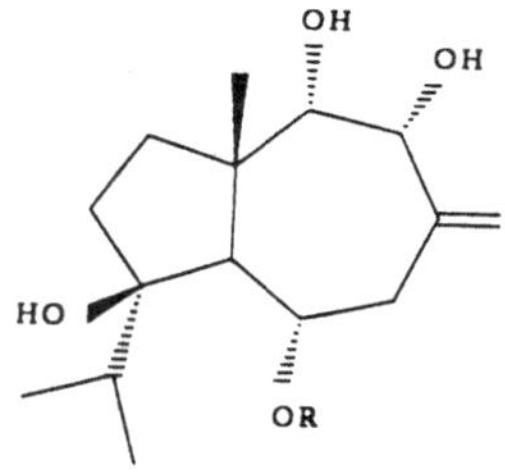

(230)   R = iVal
(231)   R = Ang

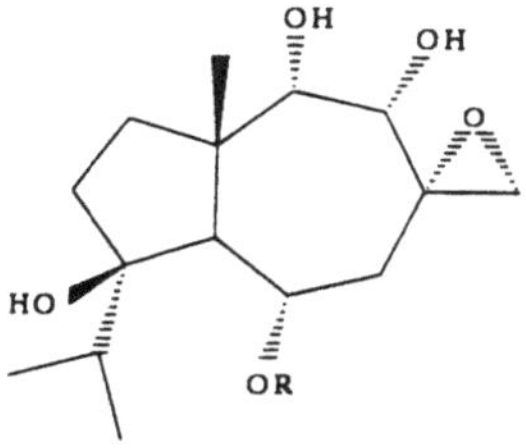

(232)   R = iVal

^{1}H NMR spectrum showed signals for an angelate, an isopropyl group (δ 0.95 and 1.02, each 3H, s), an angular methyl (δ 1.05, 3H, s), and a methyl group attached to a carbon with an oxygen group (δ 1.55, 3H, s). Finally X-ray analysis showed it to be felikiol 3-angelate.

Hydrolysis of (225) gave alcohol (155), $C_{15}H_{26}O_3$. One of the three oxygens was that of a carbonyl group (1700 cm^{-1}) while the other two were those of two tertiary hydroxy groups (3590 cm^{-1}) as no protons geminal to an oxygen were observed in the ^{1}H NMR spectrum. Reduction of (155) with sodium borohydride afforded two epimeric alcohols (156) and (157), only one of which, (157), formed an acetonide (158). The new compounds (227) and (228) had the same skeleton with the difference that in one an angelic and in the other an epoxyangelic acid esterified a tertiary hydroxy group. The structure of (228) (webiol epoxyangelate) was confirmed by X-ray analysis.

When the secondary hydroxyl at C-2 in isolancerotriol 5-isovalerate (118) was oxidized with pyridinium dichromate (90), compounds (167), (168) and (169) were obtained. The structures of these products were assigned on the basis of their ^{1}H NMR spectra and accounted for in the light of HERZ et al.'s work (91) on the oxidation of allylic alcohols. The stereochemistry assigned to C-2 of (118) was based on the assumption

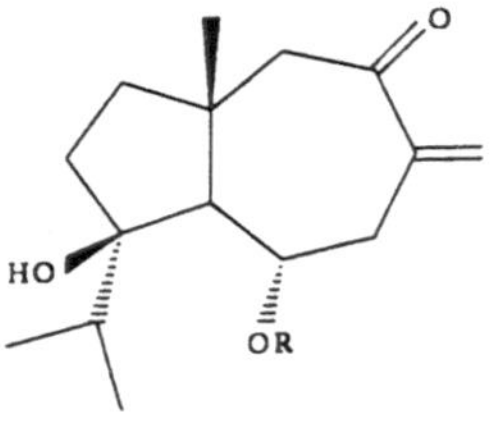

(167) 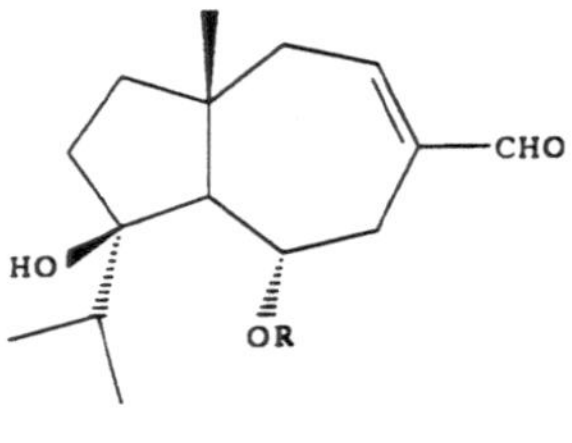   (168)

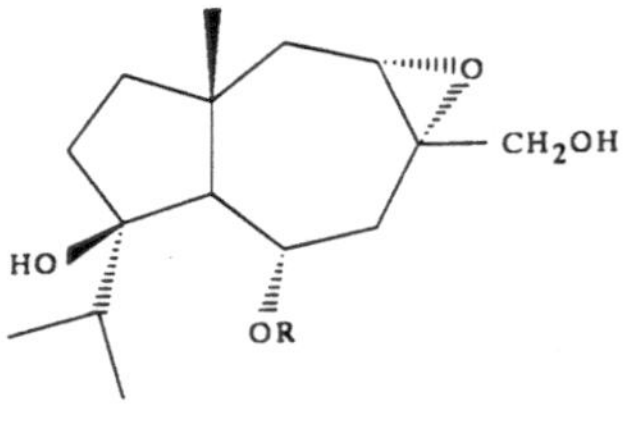

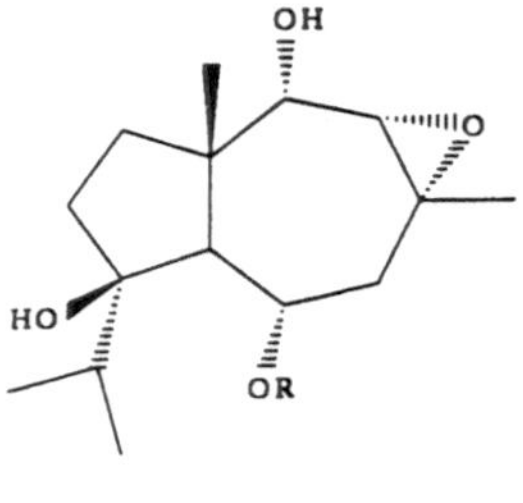

(169)   R = iVal   (170)

that it is derived from epoxyjaeschkeanadiol isovalerate (124), also isolated from this plant (55), by enzymatic cleavage of the oxirane ring with formation and neutralization of a carbocation at C-3, loss of a hydrogen from C-15 and formation of a C-3, C-15 double bond (Chart 14).

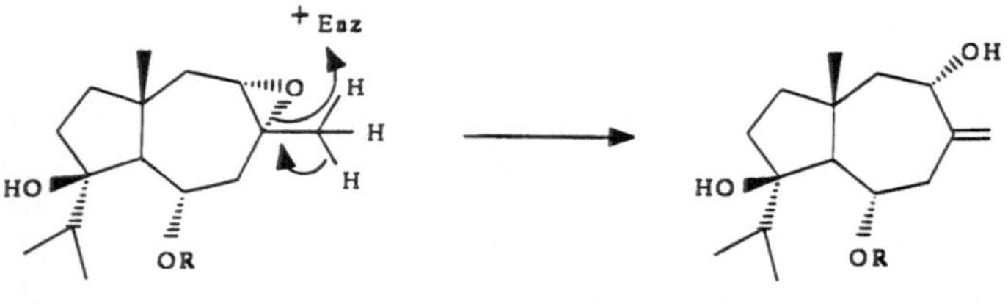

R = iVal

(124)   (118)

Chart 14. Biogenesis of lancerotriol 5-isovalerate

Isolancerotetrol 5-isovalerate (230),. $C_{20}H_{24}O_5$, was also isolated from this species. Two oxygens were part of an ester group and another belonged to a tertiary hydroxyl. The other two oxygens had to be part of two secondary hydroxyls with signals at δ3.68 and 4.31 mutually coupled ($J = 3.0$ Hz), and therefore assigned to C-1 and C-2. This was confirmed chemically as (230) formed an acetonide. The hydrogen geminal to an ester appeared as a double doublet at δ 5.32 ($J = 10.4$ and 2.0 Hz), thus establishing that the esterified secondary hydroxyl was equatorial, located between a methylene and methine group and had to be α-oriented at C-5. In all carotane sesquiterpenes from *Ferula* species which are hydroxylated at C-5 this group is α. Biogenetically, (230) may be derived from (118) by hydroxylation or from lapiferol isovalerate (179) by enzymatic cleavage of the oxirane ring with elimination of one of the hydrogens of the C-15 methyl group. Epoxyisolancerotetrol 5-isovalerate (232) was also obtained from this plant and proved identical with the *m*-chloroperbenzoic acid epoxidation product of (230).

A recent extensive study by GONZALEZ *et al.* (*80*) of the *Ferula linkii-TF* collection resulted in the isolation of sixteen sesquiterpenes with carotane skeletons: felikiol 3-angelate (225), webiol angelate (227), jaeschkeanadiol *p*-anisate (88), linkiol (221), webiol epoxyangelate (228), lapidol (200), jaeschkeanadiol angelate (89), jaeschkeanadiol veratrate (90), lapidol isobutyrate (201), lapidol 2-methylbutyrate (202), lapidol *p*-anisate (203), lapidol *p*-hydroxybenzoate (204), ferulinkiol 1-hydroxy-5(2-methylbutyrate) (171), jaeschkeanadiol isovalerate (91), jaeschkeanadiol 2-methylbutyrate (93), lapidin (205) and the humulene, 1α,10β-epoxy-3-deoxyjuniferol-*p*-anisate (183).

The ^{1}H NMR spectra of (201–204) were similar to that of lapidol (200) (*92*) differing only in the nature of the ester group on C-5, also in evidence in the mass spectra. Moreover, methylation of (204) afforded (203).

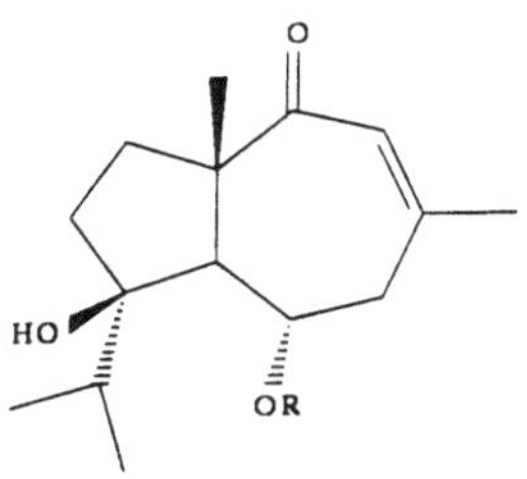

(201) R = iBut
(202) R = 2-MeBut
(203) R = *p*-Anis
(204) R = *p*-HyBz

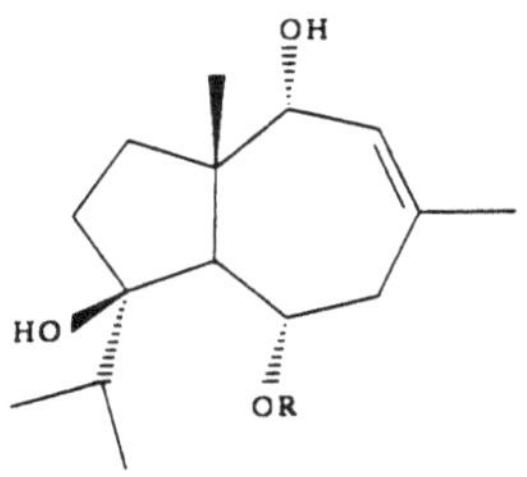

(171) R = 2-MeBut

A fifth new substance was ferulinkiol 1-hydroxy-5-(2-methylbutyrate) (171) with IR bands characteristic of hydroxy and ester groups and ^{1}H NMR signals for four methyl groups, a singlet at δ 1.71 for a methyl attached to a double bond and a doublet at δ 2.60 assigned to the hydrogen on C-6 which was coupled with the geminal hydrogen at δ 5.03 of an ester group on C-5. Conversion to lapiferol (235) as shown in Chart 15 established the complete structure and stereochemistry.

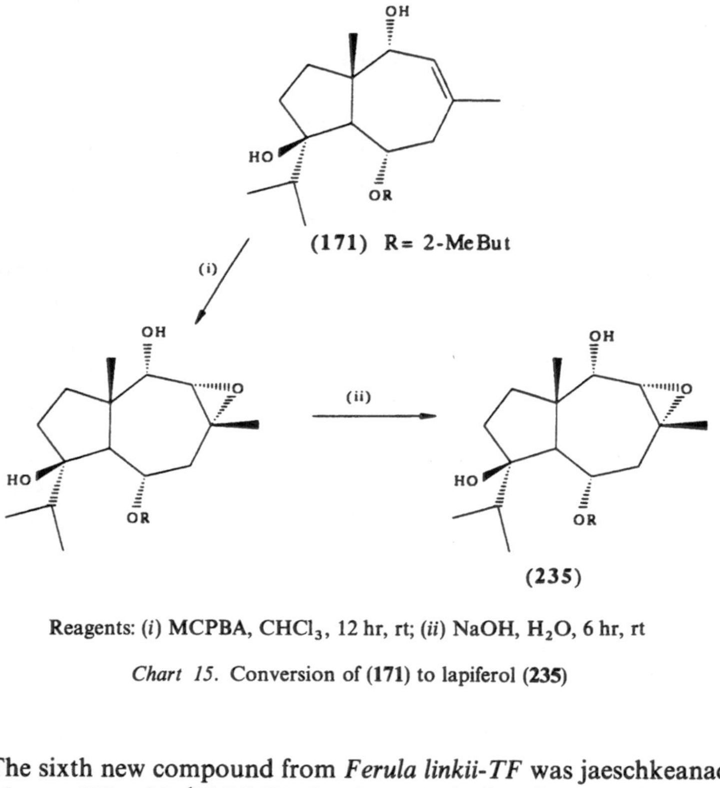

Reagents: (i) MCPBA, CHCl$_3$, 12 hr, rt; (ii) NaOH, H$_2$O, 6 hr, rt

Chart 15. Conversion of (171) to lapiferol (235)

The sixth new compound from *Ferula linkii-TF* was jaeschkeanadiol isovalerate (91) with ^{1}H NMR signals typical of an isopropyl group, an angular methyl group and a methyl group joined to a double bond. The MS showed fragments at m/z 279 [M-C$_3$H$_7$]$^+$ and 220 [M-C$_5$H$_{10}$O$_2$]$^+$ corresponding to the loss of an isopropyl radical and isovaleric acid, respectively. Analysis of the ^{1}H NMR spectrum of jaesch-keanadiol 2-methylbutyrate (93) from this collection showed that it was contaminated by jaeschkeanadiol angelate (89). Epoxidation of the mixture permitted separation of epoxyjaeschkeanadiol 2-methylbutyrate while alkaline hydrolysis of the epoxide mixture yielded epoxyjaeschkeanadiol (*55*).

Lastly, IR, mass and ¹H NMR spectra of (**234**) indicated that it was the *p*-anisyl analogue of the humulane (**57**) (*34*).

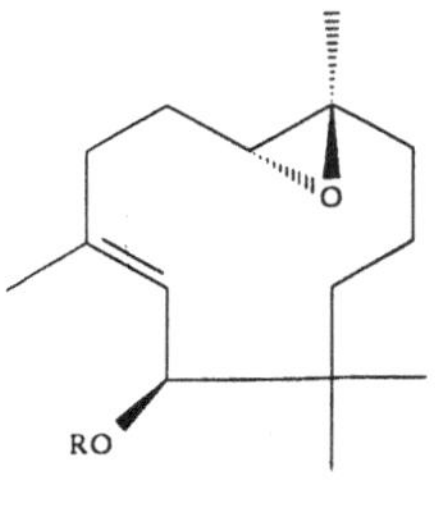

(**234**)   R = *p*-Anis

Ferula lancerottensis Parl. (*67*) yielded six new carotane sesquiterpenes: lancerodiol *p*-methoxybenzoate (**212**), lancerodiol *p*-hydroxybenzoate (**213**), lancerodiol (**214**), epoxyjaeschkeanadiol *p*-hydroxybenzoate (**126**), lancerotriol *p*-hydroxybenzoate (**237**), linkitriol *p*-methoxybenzoate (**222**) and the already known jaeschkeanadiol *p*-hydroxybenzoate (**87**).

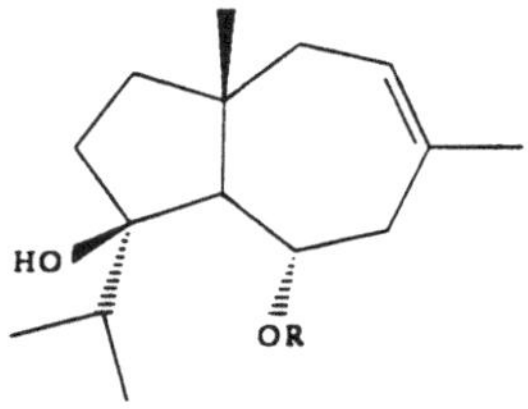

(**87**)   R = p-HyBz

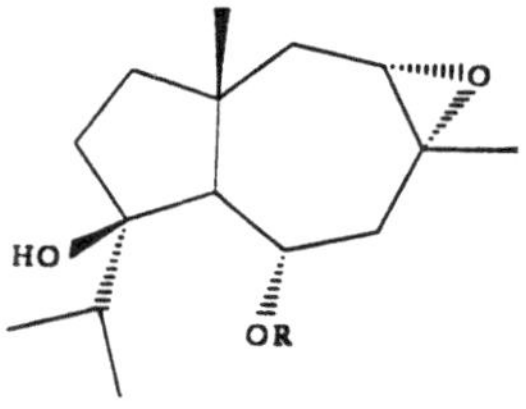

(**126**)   R = *p*-HyBz

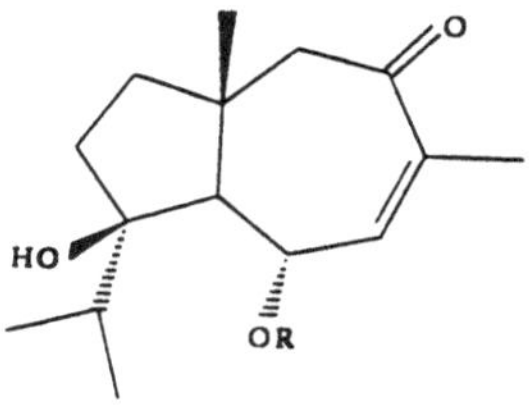

(**212**)   R = *p*-MetBz
(**213**)   R = *p*-HyBz
(**214**)   R = H

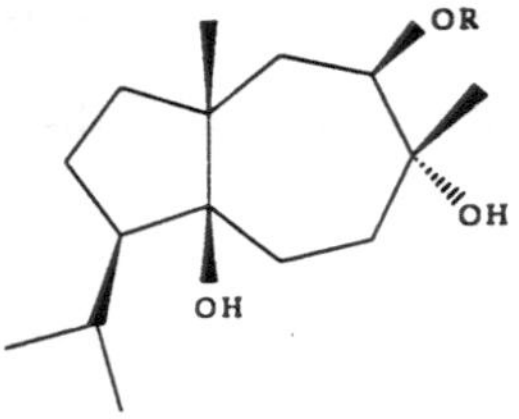

(222) R = *p*-MetBz 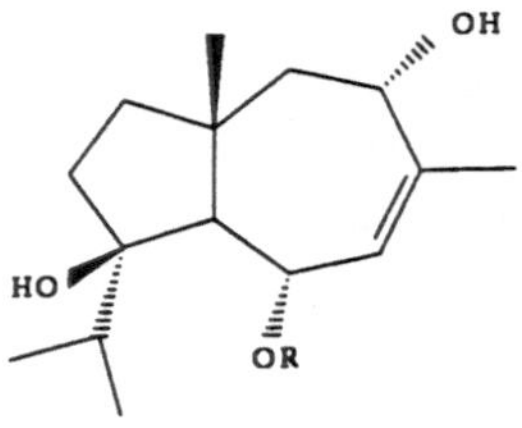   (237) R = *p*-HyBz

Compound (212) showed IR absorbances for an aromatic ester and a conjugated ketone. Its ^{1}H NMR signals were characteristic of an angular methyl group, an isopropyl group, a methyl group over a double bond, a vinyl hydrogen β to a ketone and the geminal proton of an esterified alcohol group. There were also signals for a pair of doublets (each 2H) typical of *ortho* hydrogens in an aromatic ring. The chemical transformations illustrated in Chart 16 confirmed the structure of (212).

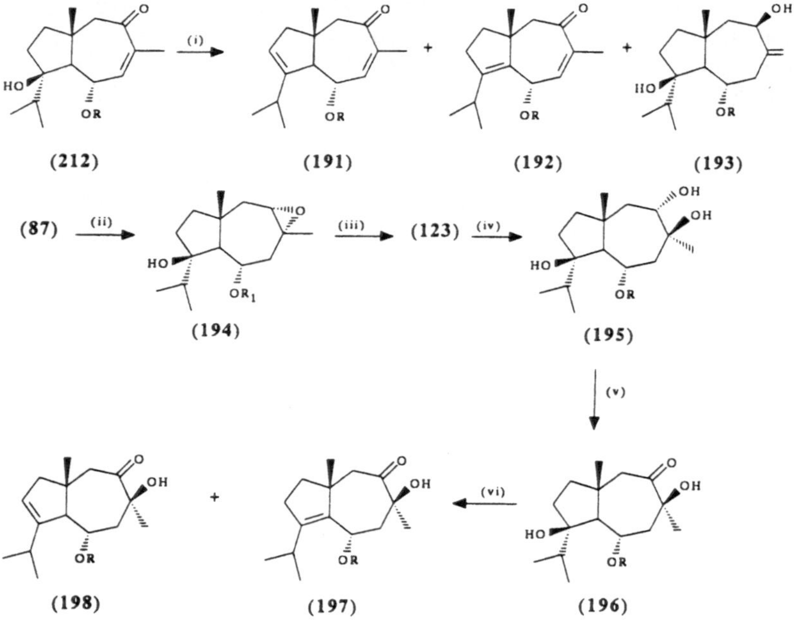

(212)   (191)   (192)   (193)

(87) → (194) → (123) → (195)

(198) + (197) ← (196)

R = *p*-Anis, R$_1$ = *p*-HyBz

Reagents: (*i*) SOCl$_2$, C$_5$H$_5$N, 5 min, 0°; (*ii*) MCPBA, CHCl$_3$, 45 min, rt; (*iii*) CH$_2$N$_2$, ether; (*iv*) aqueous 3% HClO$_4$, 45 hr, rt; (*v*) pyridinium dichromate, CH$_2$Cl$_2$, 3 hr, rt; (*vi*) SOCl$_2$, C$_5$H$_5$N, 5 min, 0°

Chart 16. Chemical transformations of lancerodiol *p*-methoxybenzoate

The most abundant substance isolated from *F. lancerottensis* was lancerodiol *p*-hydroxybenzoate (213), with a ¹H NMR spectrum similar to that of (212) except for the lack of a signal for the methoxy group. Treatment of (213) with diazomethane gave (212). Alkaline hydrolysis of (212) afforded alcohol (214) (lancerodiol). Lancerotriol *p*-hydroxybenzoate (237) was the most polar of the compounds from this species. The ¹H NMR spectrum of (222) was similar to that of linkiol (221). A probable biosynthetic pathway of jaeschkeanadiol (73) has been described (43). In *F. lancerottensis* epoxidation of jaeschkeanadiol *p*-hydroxybenzoate (87) must give ester (194). Enzymatic cleavage of the oxirane ring with elimination of one of the hydrogens over C-4 would afford lancerotriol *p*-hydroxybenzoate (237) which could undergo oxidation to give the lancerodiol ester (213) (Chart 17).

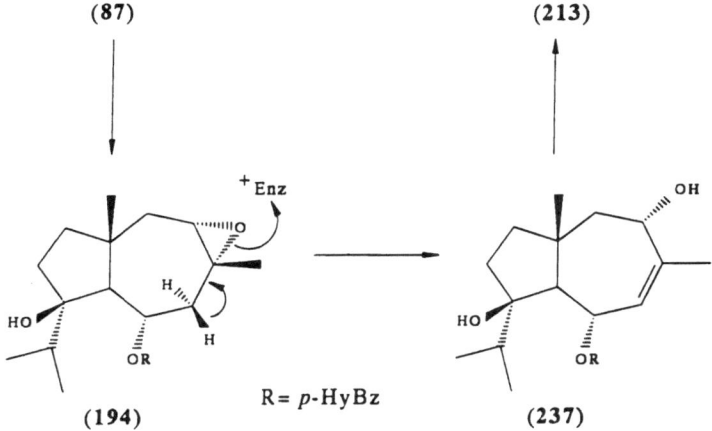

Chart 17. Biogenesis of lancerodiol *p*-hydroxybenzoate

Pallinin (172), ferulinkiol 1-angelate-5-*p*-hydroxybenzoate (173), ferulinkiol 1-angelate-5-*p*-anisate (174), lapiferol (235), lapiferin (236), lapidin (205), lapidol vanillate (206), felikiol 3-angelate (225), felikiol (226), webiol 3-angelate (227) and the humulenes (56), (57) and (58) (see page 11) were all isolated by GONZALEZ *et al.* from *Ferula latipinna* Santos (34, 80).

The new compound (206) was a derivative of lapidol. Aromatic acyl absorptions were observed in its IR spectrum and ¹H, ¹³C NMR and MS confirmed the presence of a vanillate group. Alkaline hydrolysis of (236) gave lapidol (200) and vanillic acid.

Ferula species such as *F. lancerottensis*, *F. linkii-TF*, *F. linkii-GC*, *F. latipinna*, *F. tingitana* and *F. communis*, all of which are members of

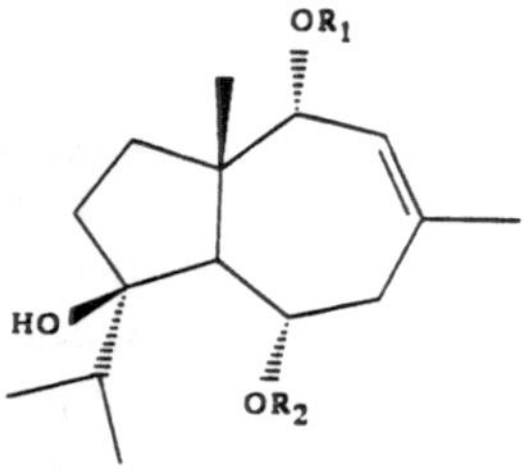

(172) $R_1 = R_2 = Ang$
(173) $R_1 = Ang$, $R_2 = p$HyBz
(174) $R_1 = Ang$, $R_2 = p$-Anis

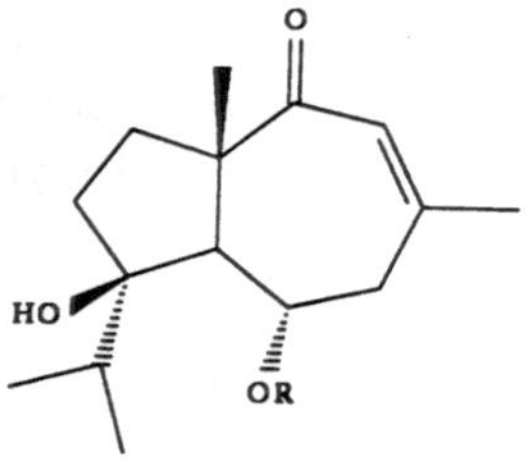

(205) R = Ang
(206) R = Van

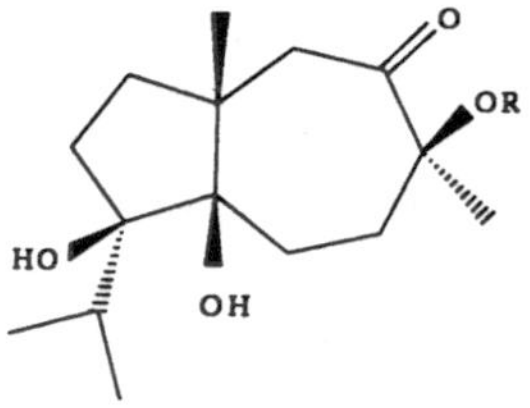

(225) R = Ang
(226) R = H

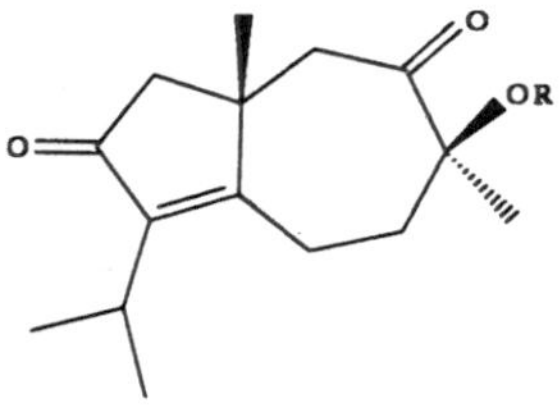

(227) R = Ang

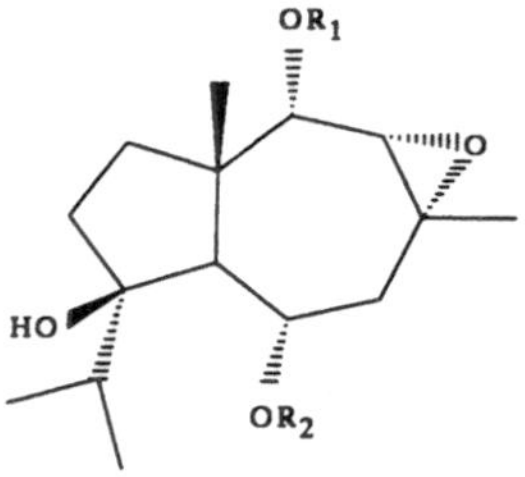

(235) $R_1 = R_2 = H$
(236) $R_1 = Ac$, $R_2 = Ang$

subgenus Euferula (Poiss.) Korovin, are of interest as they elaborate daucane derivatives with both *cis* and *trans* ring fusion. The biogenetic pathway shown in Chart 18 has been proposed by GONZALEZ *et al.* (*93*) to account for their formation.

The distribution and structures of bicyclic sesquiterpenes from *Ferula* species are shown in Table 2 and Chart 19, respectively.

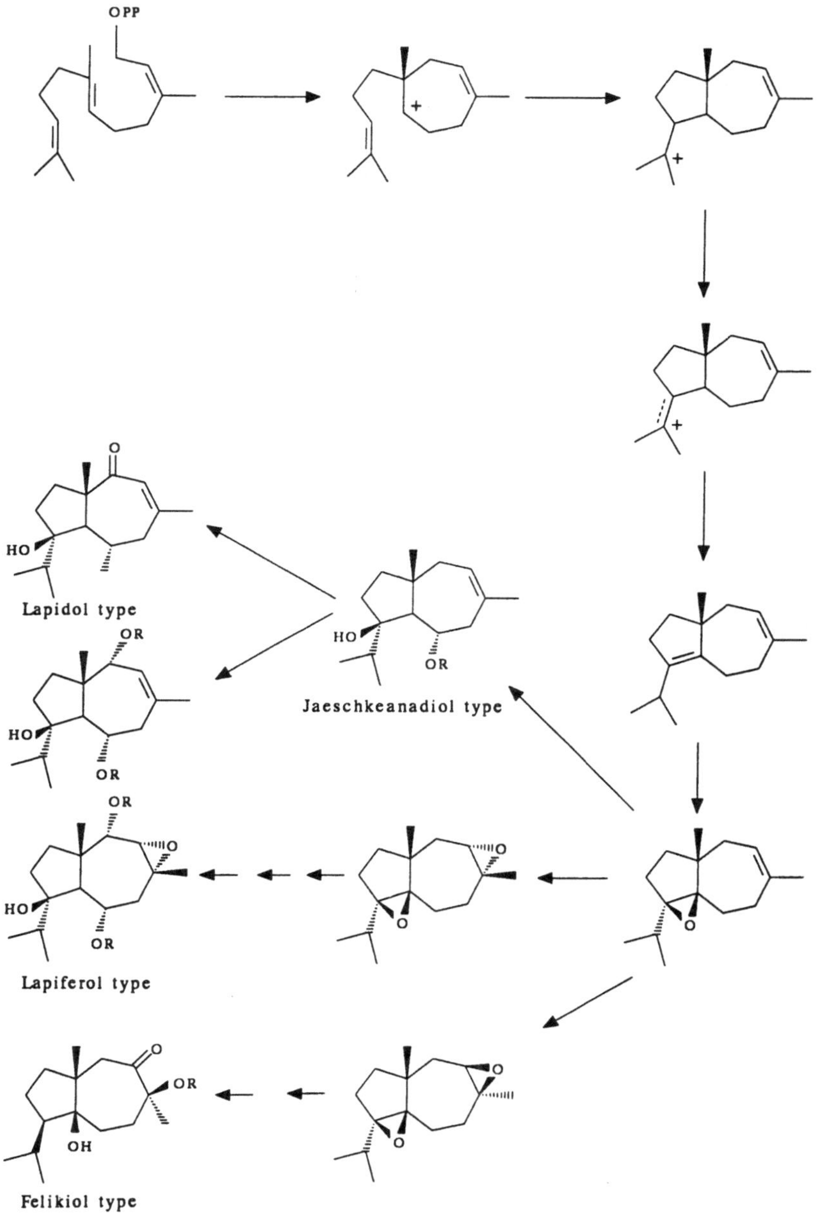

Chart 18. Biogenesis of daucane derivatives with *cis* and *trans* ring fusion

Table 2. *Bicyclic Sesquiterpenes Found in Ferula Species*

Structure number	Name of compound	Formula	Plant source	References
(59)	10-Epijunenol	$C_{15}H_{26}O$	*Ferula galbaniflua*	(35)
(70)	Xeroferin	$C_{23}H_{32}O_5$	*F. xeromorpha*	(42)
(73)	Jaeschkeanadiol	$C_{15}H_{26}O_2$	*F. jaeschkeana* Vatke	(43, 125)
(87)	Ferutinin (jaeschkeanadiol *p*-hydroxybenzoate)	$C_{22}H_{30}O_4$	*F. elaeochytris*	(54)
			F. akitschensis	(119)
			F. jaeschkeana Vatke	(125–127)
			F. ovina	(120)
			F. pallida	(121)
			F. sinaica	(122)
			F. rigidula	(134)
			F. tenuisecta	(133)
			F. lancerottensis	(67)
			F. communis	(132)
			F. orientalis	(23)
			F. elaeochytris	(54)
(88)	Jaeschkeanadiol *p*-methoxybenzoate (Ferutidin)	$C_{23}H_{32}O_4$	*F. linkii-GC*	(56)
			F. linkii-TF	(79)
			F. communis	(63)
			F. elaeochytris	(54)
			F. rigidula	(134)
			F. kuhistanica	(135)
			F. sinaica	(147)
			F. sinaica	(122)
(89)	Jaeschkeanadiol angelate	$C_{20}H_{32}O_3$	*F. linkii-TF*	(79)
			F. elaeochytris	(54)
			F. jaeschkeana	(125)
(90)	Jaeschkeanadiol veratrate	$C_{24}H_{34}O_5$	*F. linkii-TF*	(79)

(91)	Jaeschkeanadiol iso-valerate	$C_{20}H_{34}O_3$	*F. linkii-TF*	(79)
(93)	Jaeschkeanadiol 2-methyl-butyrate	$C_{20}H_{34}O_3$	*F. linkii-TF*	(79)
(94)	Ferutin	$C_{23}H_{32}O_5$	*F. akitschensis*	(119)
			F. ovina	(120)
			F. pallida	(121)
			F. sinaica	(122)
			F. tenuisecta	(123, 124)
			F. soongarica	(104)
(95)	Akiferidin	$C_{22}H_{30}O_5$	*F. akitschensis*	(119)
(96)	Akiferin	$C_{24}H_{34}O_5$	*F. akitschensis*	(128)
(97)	Teferidine	$C_{22}H_{30}O_3$	*F. elaeochytris*	(54)
			F. pallida	(121)
			F. tenuisecta	(133)
			F. rigidula	(134)
			F. jaeschkeana	(125)
(98)	Jaeschkeanadiol 5α-(3-methoxy-4-hydroxy-benzoate)(Teferin)	$C_{23}H_{32}O_5$	*F. elaeochytris*	(54)
			F. tenuisecta	(136)
			F. orientalis	(23)
			F. soongarica	(104)
			F. pallida	(121)
			F. jaeschkeana	(126, 127)
			F. rigidula	(134)
(99)	Jaeschkeanadiol salicylate	$C_{22}H_{30}O_4$	*F. elaeochytris*	(54)
			F. jaeschkeana	(125)
			F. rigidula	(134)
(100)	Jaeschkeanidin	$C_{23}H_{30}O_5$	*F. jaeschkeana*	(125)
(101)	Ferutinianin	$C_{22}H_{30}O_5$	*F. jaeschkeana*	(125)
(102)	Palliferidin	$C_{25}H_{36}O_6$	*F. pallida*	(144)
(113)	Jaeschkeanadiol benzoate	$C_{22}H_{30}O_3$	*F. sinaica*	(122)
			F. rigidula	(134)

Table 2 (continued)

Structure number	Name of compound	Formula	Plant source	References
(114)	Jaeschkeanadiol t-cinnamate	$C_{24}H_{32}O_3$	F. rigidula	(134)
(115)	Feruginidin	$C_{22}H_{30}O_5$	F. jaeschkeana Vatke	(51)
(117)	Ferugin	$C_{22}H_{30}O_5$	F. jaeschkeana Vatke	(51)
(118)	Isolancerotriol 5-isovalerate	$C_{20}H_{34}O_4$	F. linkii-GC	(72)
(119)	Isolancerotriol	$C_{15}H_{26}O_3$	F. sinaica	(122)
(120)	Isolancerotriol 6-vanillate	$C_{23}H_{32}O_6$	F. rigidula	(134)
(121)	2,3-Epoxyjaeschkeanadiol 5α-vanillate	$C_{23}H_{32}O_6$	F. jaeschkeana Vatke	(52)
			F. orientalis	(23)
(122)	2,3-Epoxyjaeschkeanadiol	$C_{15}H_{26}O_3$	F. linkii-GC	(55)
(123)	Epoxyjaeschkeanadiol p-methoxybenzoate	$C_{23}H_{32}O_5$	F. linkii-GC	(55)
			F. rigidula	(134)
(124)	Epoxyjaeschkeanadiol isovalerate	$C_{20}H_{34}O_4$	F. linkii-GC	(55)
	Epoxyjaeschkeanadiol veratrate	$C_{24}H_{34}O_6$	F. linkii-GC	(55)
(126)	Epoxyjaeschkeanadiol p-hydroxybenzoate (Jaschkeanin)	$C_{22}H_{30}O_5$	F. lancerottensis Parl	(67)
			F. jaeschkeana	(125)
			F. orientalis	(23)
(127)	Epoxyjaeschkeanadiol t-cinnamate	$C_{24}H_{32}O_4$	F. rigidula	(134)
(128)	2β-Hydroxy-3,4-epoxyjaeschkeanadiol	$C_{22}H_{30}O_5$	F. jaeschkeana Vatke	(52)
(129)	Akitschenol (= 153)	$C_{15}H_{26}O_3$	F. akitschensis	(118)
(131)	Akichenol	$C_{22}H_{30}O_5$	F. jaeschkeana Vatke	(53)
(132)	5α-O-p-hydroxybenzoate Akichenol 9β-O-angelate	$C_{20}H_{34}O_5$	F. jaeschkeana Vatke	(53)

No.	Compound	Formula	Species	Ref.
(138)	Jaeschkeanadiol 8β-hydroxy-5-p-anisate	$C_{23}H_{32}O_5$	*F. sinaica*	(68, 69)
(140)	Unnamed	$C_{24}H_{32}O_5$	*F. communis*	(63)
(141)	Akiferidinin	$C_{27}H_{36}O_7$	*F. akitschensis*	(118, 149)
(148)	Jaeschferin	$C_{23}H_{32}O_6$	*F. jaeschkeana*	(58)
(149)	Ferutinianin angelate	$C_{27}H_{36}O_7$	*F. jaeschkeana*	(125)
(150)	9β-Hydroxyjaeschkeanadiol	$C_{15}H_{26}O_3$	*F. jaeschkeana*	(127)
(153)	Akichenol (= **129**)	$C_{15}H_{26}O_3$	*F. jaeschkeana*	(53)
(154)	Ferujaesenol	$C_{15}H_{24}O_2$	*F. jaeschkeana* Vatke	(53)
(159)	Feruone	$C_{15}H_{24}O_3$	*F. jaeschkeana* Vatke	(57)
(160)	Ferutriol 5α-(p-hydroxybenzoate)	$C_{22}H_{30}O_5$	*F. jaeschkeana* Vatke	(57)
			F. sinaica	(122)
(161)	Tingitanol	$C_{25}H_{38}O_6$	*F. tingitana* L.	(60)
(162)	Deoxodshydrolaserpitine	$C_{25}H_{38}O_6$	*F. tingitana* L.	(62)
(163)	Tingitanol acetate	$C_{27}H_{40}O_7$	*F. tingitana* L.	(60)
(164)	Deoxodehydro-laserpitine acetate	$C_{27}H_{40}O_7$	*F. tingitana* L.	(62)
(165)	4β-Hydroxy-6α(p-hydroxybenzoyloxy)-10α-angeloxydauc-7-ene	$C_{27}H_{36}O_6$	*F. tingitana* L.	(62)
			F. sinaica	(122)
(166)	10β-Hydroxy-5α-p-hydroxybenzoyl-1α-angeloyloxydauc 2-ene	$C_{28}H_{38}O_6$	*F. communis* subsp. *communis*	(63)
(171)	Ferulinkiol 1α-hydroxy-5(2-methylbutyrate)	$C_{20}H_{34}O_4$	*F. linkii-TF*	(79)
(172)	Pallinin (8-deoxytingitanol)	$C_{23}H_{32}O_5$	*F. latipinna* Santos	(33, 79)
			F. pallida	(146)
(173)	Ferulinkiol-1-angelate-5-p-hydroxybenzoate	$C_{27}H_{36}O_6$	*F. latipinna* Santos	(33, 79)
(174)	Ferulinkiol-1-angelate-5-p-anisate	$C_{28}H_{38}O_6$	*F. latipinna* Santos	(33, 79)
(175)	Unnamed	$C_{23}H_{32}O_5$	*F. communis* subsp. *communis*	(63)
(176)	Unnamed	$C_{25}H_{34}O_6$	*F. communis* subsp. *communis*	(63)
(177)	Unnamed	$C_{28}H_{38}O_6$	*F. communis* subsp. *communis*	(63)
(178)	Unnamed	$C_{28}H_{38}O_6$	*F. communis* subsp. *communis*	(63)

Table 2 (continued)

Structure number	Name of compound	Formula	Plant source	References
(179)	Unnamed	$C_{29}H_{40}O_7$	F. communis subsp. communis	(63)
(180)	Unnamed	$C_{24}H_{32}O_5$	F. sinaica	(68, 69)
(181)	Akitschenin	$C_{27}H_{36}O_6$	F. akitschensis	(118)
(182)	Unnamed	$C_{24}H_{32}O_5$	F. communis subsp. communis	(63)
(183)	Unnamed	$C_{25}H_{34}O_6$	F. communis subsp. communis	(63)
(184)	5α-(4-Hydroxybenzoyl)-9β-angeloxyjaeschkeanadiol	$C_{27}H_{36}O_6$	F. jaeschkeana Vatke	(127)
(185)	Isofercomin	$C_{23}H_{30}O_5$	F. sinaica	(68, 69)
(186)	Isolancerotriol-2-epi-5-p-hydroxybenzoate	$C_{22}H_{30}O_5$	F. sinaica	(69)
(187)	3-5α,10β-Trihydroxy-2α-p-hydroxybenzoyloxy-p-hydroxybenzoyloxy-daucane	$C_{22}H_{32}O_6$	F. sinaica	(69, 122)
(188)	Unnamed	$C_{22}H_{32}O_6$	F. sinaica	(69, 122)
(189)	5α,10β-Dihydroxydauc-2-ene-1-one	$C_{15}H_{24}O_3$	F. sinaica	(69)
(190)	5α-p-Hydroxybenzoyloxy-10-hydroxy-dauc-2-ene-1-one	$C_{22}H_{28}O_5$	F. sinaica	(69)
(199)	5α,8β-Diangeloyloxy-10β-hydroxydauc-2-ene-1-ene	$C_{22}H_{36}O_6$	F. sinaica	(69)
(200)	Lapidol	$C_{15}H_{24}O_3$	F. linkii-TF	(79)
(201)	Lapidol isobutyrate	$C_{19}H_{30}O_4$	F. linkii-TF	(79)
(202)	Lapidol 2-methylbutyrate	$C_{20}H_{32}O_4$	F. linkii-TF	(79)
(203)	Lapidol p-anisate	$C_{23}H_{30}O_5$	F. linkii-TF	(79)
(204)	Lapidol p-hydroxybenzoate	$C_{22}H_{28}O_5$	F. linkii-TF	(79)
(205)	Lapidin	$C_{20}H_{30}O_4$	F. latipinna Santos	(33, 79)
			F. linkii-TF	(79)
			F. lapidosa	(92)
(206)	Lapidol vanillate	$C_{23}H_{30}O_6$	F. latipinna Santos	(33, 79)
(207)	Palliferin	$C_{25}H_{34}O_7$	F. pallida	(145)
(208)	Palliferinin	$C_{24}H_{30}O_7$	F. pallida	(145)
(209)	Fercomin = (218)	$C_{23}H_{30}O_5$	F. communis subsp. communis	(64)

	Name	Formula	Species	Ref.
(210)	Lancerodiol 8β-hydroxy-5-p-hydroxybenzoate	$C_{22}H_{28}O_6$	*F. sinaica*	(69)
(211)	Lancerotol veratrate	$C_{24}H_{32}O_6$	*F. linkii-GC*	(56)
(212)	Lancerodiol p-methoxy-benzoate	$C_{23}H_{30}O_5$	*F. lancerottensis* Parl	(67)
(213)	Lancerodiol p-hydroxy-benzoate	$C_{22}H_{28}O_5$	*F. lancerottensis* Parl	(67)
			F. jaeschkeana Vatke	(53)
			F. rigidula	(134)
			F. orientalis	(23)
(214)	Lancerodiol	$C_{15}H_{24}O_3$	*F. lancerottensis* Parl	(67)
(215)	Lancerodiol vanillate	$C_{23}H_{30}O_6$	*F. orientalis*	(23)
			F. rigidula	(134)
(216)	Lancerodiol p-coumarate	$C_{24}H_{30}O_5$	*F. rigidula*	(134)
(217)	Lancerodiol t-cinnamate	$C_{24}H_{30}O_4$	*F. rigidula*	(134)
(218)	8-Keto-6-hydroxy-1-p-anisoyloxy-dauc-2-ene = Fercomin (209)	$C_{23}H_{30}O_5$	*F. communis* subsp. *communis*	(64)
			F. tingitana	(62)
(219)	Rel-(3R,3aR,8S,8aR)-3a-hydroxy-1-oxo-6-dauc-8-yl-p-anisate	$C_{23}H_{30}O_5$	*F. communis*	(131)
(220)	Fercolide	$C_{23}H_{28}O_5$	*F. communis* subsp. *communis*	(64)
(221)	Linkiol	$C_{20}H_{34}O_4$	*F. linkii-GC*	(82)
			F. linkii-TF	(79)
(222)	Linkitriol p-methoxy-benzoate	$C_{23}H_{34}O_5$	*F. lancerottensis* Parl	(67)
(223)	Carotdiol acetate	$C_{17}H_{26}O_3$	*F. linkii-GC*	(85)
(224)	Carotdiol veratrate	$C_{24}H_{34}O_5$	*F. linkii-GC*	(85)
(225)	Felikiol 3-angelate	$C_{20}H_{32}O_4$	*F. linkill-GC*	(56, 70)
			F. tenuisecta	(148)
(226)	Felikiol	$C_{15}H_{26}O_3$	*F. latipinna* Santos	(34, 79)
(227)	Webiol angelate	$C_{20}H_{30}O_4$	*F. latipinna* Santos	(33, 79)
			F. linkii-GC	(56)
			F. latipinna Santos	(33, 79)

Table 2 (*continued*)

Structure number	Name of compound	Formula	Plant source	References
(228)	Webiol epoxyangelate	$C_{20}H_{30}O_5$	*F. linkii-GC*	(56)
			F. linkii-TF	(79)
			F. linkii-GC	(56)
(229)	Ferutriol 5-isovalerate	$C_{20}H_{34}O_4$		
(230)	Isolancerotetrol 5-isovalerate	$C_{20}H_{34}O_5$	*F. linkii-GC*	(70)
(231)	Isolancerotetrol 5-angelate	$C_{20}H_{32}O_5$	*F. linkii-GC*	(70)
(232)	Epoxylanceroterol 5-isovalerate	$C_{20}H_{34}O_6$	*F. linkii-GC*	(70)
(233)	4β,8β-Dihydro-6α-vanilloyl-oxydauc-9-ene	$C_{23}H_{32}O_6$	*F. rigidula*	(134)
(234)	Humulene 1α,10β-epoxy-3-deoxyjuniferol-*p*-anisate	$C_{23}H_{32}O_4$	*F. linkii-TF*	(79)
(235)	Lapiferol	$C_{15}H_{26}O_4$	*F. latipinna* Santos	(33, 79)
(236)	Lapiferin	$C_{22}H_{34}O_6$	*F. latipinna* Santos	(33, 79)
			F. lapidosa	(141)
(237)	Lancerotriol *p*-hydroxy-benzoate	$C_{22}H_{30}O_5$	*F. lancerottensis* Parl	(67)
(238)	Lancerotriol 6-vanillate	$C_{23}H_{32}O_6$	*F. rigidula*	(134)
(294)	Akiferinin	$C_{24}H_{34}O_6$	*F. akitschensis*	(129)
(240)	Tenuferin	$C_{23}H_{32}O_6$	*F. tenuisecta*	(130)
			F. tenuisecta	(130)
(241)	Tenuferinin	$C_{23}H_{32}O_6$	*F. tenuisecta*	(130)
(242)	Tenuferidin	$C_{22}H_{30}O_5$	*F. tenuisecta*	(130)
(243)	Rel-(1S,3R,3aS,4S,8S, 8aR)-1,8-diacetoxy-3-hydroxy-6-daucen-4-yl *p*-anisate	$C_{27}H_{36}O_8$	*F. communis*	(131)

(244)	Unnamed	$C_{25}H_{34}O_7$	*F. communis* subsp. *communis*	(63)
(245)	Unnamed	$C_{27}H_{36}O_8$	*F. communis* subsp. *communis*	(63)
(246)	Unnamed	$C_{27}H_{36}O_8$	*F. communis* subsp. *communis*	(63)
(247)	Unnamed	$C_{28}H_{38}O_9$	*F. communis* subsp. *communis*	(63)
(248)	Unnamed	$C_{23}H_{32}O_4$	*F. communis* subsp. *communis*	(63)
(249)	Unnamed	$C_{23}H_{30}O_4$	*F. communis* subsp. *communis*	(63)
(250)	Ferutinone	$C_{22}H_{28}O_5$	*F. jaeschkeana*	(125)
(251)	Lapidolinin	$C_{28}H_{38}O_{10}$	*F. lapidosa*	(137, 138)
(252)	Lapidolin	$C_{23}H_{36}O_8$	*F. lapidosa*	(137)
(253)	Lapidolidin	$C_{26}H_{36}O_9$	*F. lapidosa*	(140)
(254)	Lapiferinin	$C_{26}H_{36}O_8$	*F. lapidosa*	(139)
(255)	Microferin	$C_{22}H_{28}O_3$	*F. microcarpa*	(142)
(256)	Microferinin	$C_{23}H_{30}O_4$	*F. microcarpa*	(142)
(257)	Pallidin	$C_{25}H_{38}O_4$	*F. pallida*	(143)
(258)	Taulin	$C_{26}H_{34}O_7$	*F. pallida*	(121)
(259)	Taufenin	$C_{20}H_{34}O_4$	*F. pallida*	(121)
(260)	Taufedin	$C_{25}H_{38}O_7$	*F. pallida*	(121)
(261)	Tauferin	$C_{20}H_{34}O_4$	*F. pallida*	(121)
(262)	14-Hydroxyvaginatin	$C_{20}H_{30}O_5$	*F. sinaica*	(122)
(263)	14-*p*-Anisoyloxydauc-4,8-diene	$C_{23}H_{30}O_3$	*F. tingitana*	(62)
(264)	Torosoldiangelate	$C_{25}H_{36}O_5$	*F. rigidula*	(134)
(265)	4β,8α-Dihydroxy-6α-vanilloyloxydauc-9-ene	$C_{23}H_{32}O_6$	*F. rigidula*	(134)
(266)	Daucol 13-vanillate	$C_{23}H_{32}O_6$	*F. rigidula*	(134)

Chart 19. Structures of bicyclic sesquiterpenes found in Ferula

(59) 10-Epijuneol

(70) Xeroferin; R = iVan

(73) Jaeschkeanadiol; R = H
(87) Jaeschkeanadiol p-hydroxybenzoate (Ferutinin);
      R = p-HyBz
(88) Jaeschkeanadiol p-methoxybenzoate (Ferutidin);
      R = p-Anis
(89) Jaeschkeanadiol angelate; R = Ang
(90) Jaeschkeanadiol veratrate; R = Ver
(91) Jaeschkeanadiol isovalerate; R = iVal
(93) Jaeschkeanadiol 2-methylbutyrate; R = 2-MeBu
(94) Ferutin; R = iVan
(95) Akiferidin; R = 3,4-dihydroxybenzoate
(96) Akiferin; R = Ver
(97) Teferidine; R = Bz
(98) Jaeschkeanadiol 5α-(3-methoxy-4-
      hydroxybenzoate) (Teferin); R = Van
(99) Jaeschkeanadiol salicylate; R = Salicylate
(100) Jaeschkeanidin; R = 3,4-
      methylenedioxybenzoate
(101) Ferutinianin; R = 3,4-dihydroxybenzoate
(102) Palliferidin; R = trimethoxygallicate
(113) Jaeschkeanadiol benzoate; R = Bz
(114) Jaeschkeanadiol t-cinnamate; R = t-Cinn

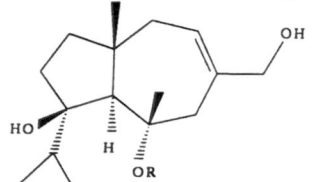

(115) Feruginidin; R = p-HyBz

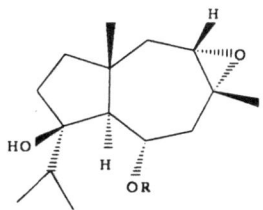

(117) Ferugin; R = *p*-HyBz
(118) Isolancerotriol 5-isovalerate; R = iVal
(119) Isolancerotriol; R = H
(120) Isolancerotriol 6-vanillate; R = Van

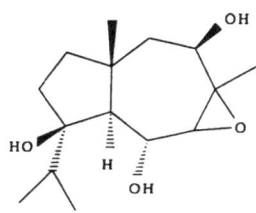

(121) 2,3-Epoxyjaeschkeanadiol 5α-vanillate;
R = Van
(122) 2,3-Epoxyjaeschkeanadiol; R = H
(123) Epoxyjaeschkeanadiol *p*-methoxybenzoate;
R = *p*-Anis
(124) Epoxyjaeschkeanadiol isovalerate; R = iVal
(125) Epoxyjaeschkeanadiol veratrate, R = Ver
(126) Epoxyjaeschkeanadiol *p*-hydroxybenzoate
(Jaeschkeanin); R = *p*-HyBz
(127) Epoxyjaeschkeanadiol *t*-cinnamate; R = *t*-Cinn

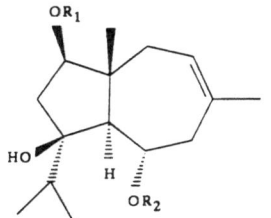

(128) 3,4-Epoxyjaeschkeanadiol 2β-hydroxy

(129) Akitschenol; $R_1 = R_2 = H$ (= **153**)
(131) Akichenol 5α-O-*p*-Hydroxybenzoate;
$R_1 = H$, $R_2 = p$-HyBz
(132) Akichenol 9β-O-angelate; $R_1 = Ang$, $R_2 = H$
(138) Jaeschkeanadiol 8β-hydroxy-5-*p*-anisate;
$R_1 = H$, $R_2 = p$-Anis
(140) Unnamed; $R_1 = Ac$, $R_2 = Bz$
(141) Akiferidinin; $R_1 = Ang$, $R_2 = 3,4$-
dihydroxybenzoate
(148) Jaeschferin; $R_1 = H$, $R_2 = Van$
(149) Ferutinianin angelate; $R_1 = Ang$, $R_2 = 3,4$-
dihydroxybenzoate
(150) 9β-Hydroxyjaeschkeanadiol $R_1 = R_2 = H$
(153) Akichenol; $R_1 = R_2 = H$ (= **129**)

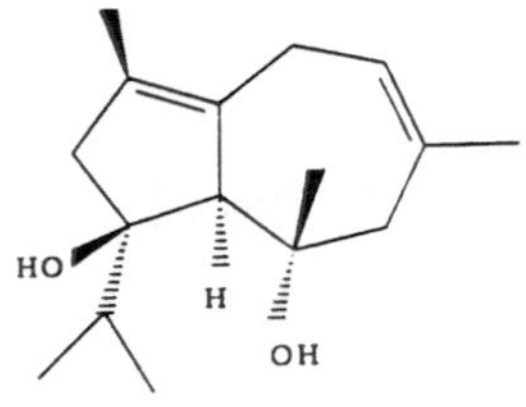

(154)   Ferujaesenol

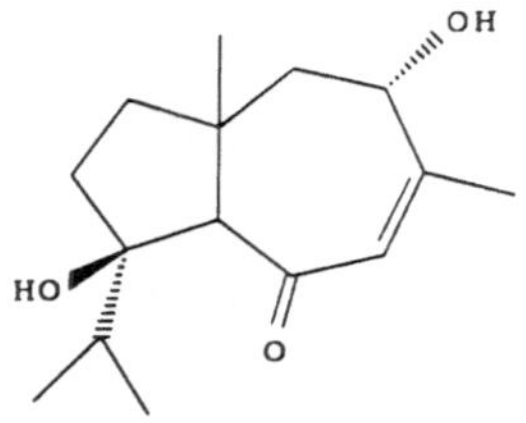

(159)   Feruone

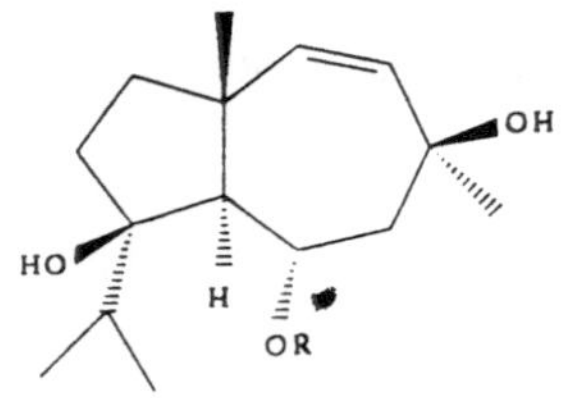

(160)   Ferutriol 5α-(p-hydroxybenzoate); R = p-HyBz

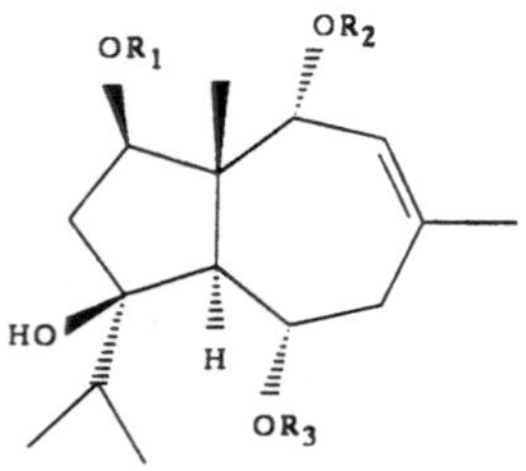

(161)   Tingitanol; R_1 = H, R_2 = R_3 = Ang
(162)   Deoxodehydrolaserpitine; R_1 = R_3 = Ang, R_2 = H
(163)   Tingitanol acetate; R_1 = Ac, R_2 = R_3 = Ang
(164)   Deoxodehydrolaserpitine acetate; R_1 = R_3 = Ang, R_2 = Ac

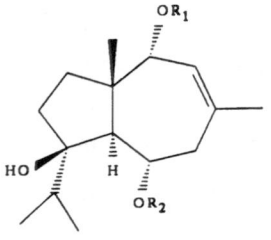

(165)   4β-Hydroxy-6α-(p-hydroxybenzoyloxy)-10α-angeloxydauc-7-ene; R_1 = Ang, R_2 = p-HyBz
(166)   10β-Hydroxy-5α-p-anisoyloxy-1α-angelyloxydauc-2-ene; R_1 = Ang, R_2 = p-Anis

(171) Ferulinkiol 1α-hydroxy-5(2-methylbutyrate);
       $R_1 = H$, $R_2 = $ 2-MeBut
(172) Pallinin (8-deoxytingitanol); $R_1 = R_2 = $ Ang
(173) Ferulinkol-1-angelate-5-*p*-hydroxybenzoate;
       $R_1 = $ Ang, $R_2 = p$-HyBz
(174) Ferulinkiol-1-angelate-5-*p*-anisate; $R_1 = $ Ang,
       $R_2 = p$-Anis
(175) Unnamed; $R_1 = H$, $R_2 = p$-Anis
(176) Unnamed; $R_1 = $ Ac, $R_2 = p$-Anis
(177) Unnamed; $R_1 = $ Ang, $R_2 = $ Bz
(178) Unnamed; $R_1 = $ Ang, $R_2 = p$-Anis
(179) Unnamed; $R_1 = $ Ang, $R_2 = $ Ver

(180) Unnamed; $R_1 = $ Bz, $R_2 = $ Ac
(181) Akichenin; $R_1 = $ Ang, $R_2 = p$-HyBz
(182) Unnamed; $R_1 = $ Ac, $R_2 = $ Bz
(183) Unnamed; $R_1 = $ Ac, $R_2 = p$-Anis

(184) Jaeschkeanadiol 5α-(4-hydroxybenzoyl)-9β-
       angelate; $R_1 = $ Ang, $R_2 = p$-HyBz

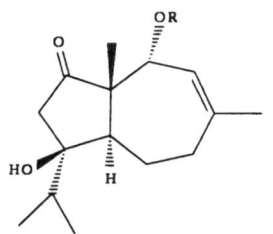

(185) Isofercomin; $R = p$-Anis

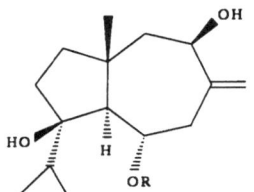

(186) Isolancerotriol 2-*epi*-5-*p*-hydroxybenzoate;
       $R = p$-HyBz

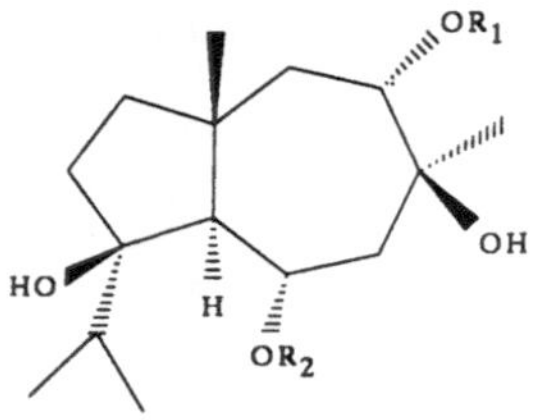

(187) 3β,5α,10β-Trihydroxy-2α-*p*-hydroxy-
benzoyloxydaucane; $R_1 = p$-HyBz, $R_2 = $ H
(188) Unnamed; $R_1 = $ H, $R_2 = p$-HyBz

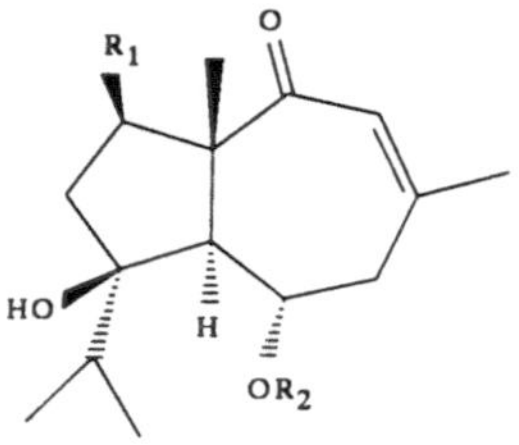

(189) 5α,10β-Dihydroxydauc-2-ene-1-one;
$R_1 = R_2 = $ H
(190) 5α-*p*-Hydroxybenzoyloxydauc-2-ene-1-one;
$R_1 = $ H, $R_2 = p$-HyBz
(199) 5α,8β-Diangeloyloxy-10β-hydroxydauc-2-ene-
1-one; $R_1 = R_2 = $ Ang
(200) Lapidol; $R_1 = R_2 = $ H
(201) Lapidol isobutyrate; $R_1 = $ H, $R_2 = $ iBut
(202) Lapidol 2-methylbutyrate; $R_1 = $ H,
$R_2 = $ 2-MeBut
(203) Lapidol *p*-anisate; $R_1 = $ H, $R_2 = p$-Anis
(204) Lapidol *p*-hydroxybenzoate; $R_1 = $ H,
$R_2 = p$-HyBz
(205) Lapidin; $R_1 = $ H, $R_2 = $ Ang
(206) Lapidol vanillate; $R_1 = $ H, $R_2 = $ Van
(207) Palliferin; $R_1 = $ H, $R_2 = $ Trimethoxygallicate
(208) Palliferinin; $R_1 = $ H, $R_2 = $ 3,4-
methylenedioxybenzoate

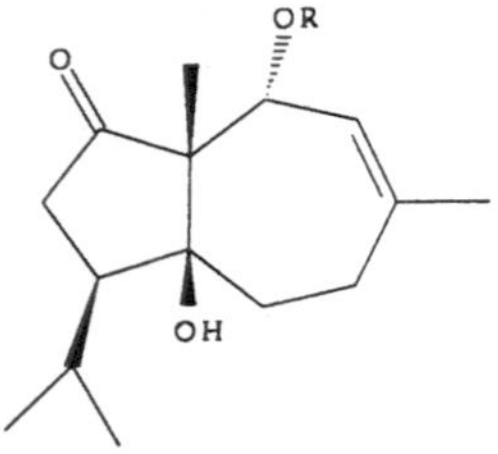

(209) Fercomin = (218); $R = p$-Anis

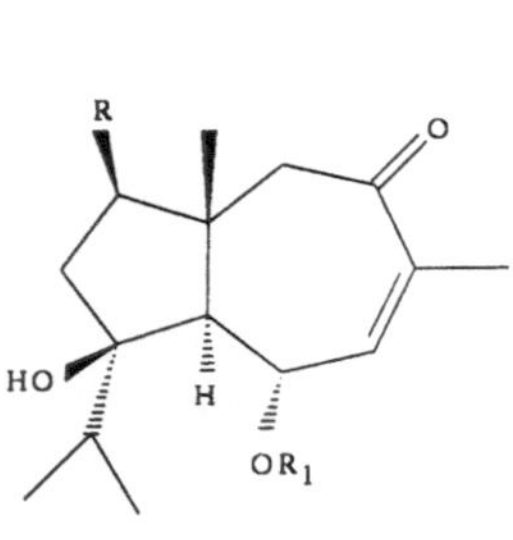

(210) Lancerodiol 8β-hydroxy-5-*p*-hydroxybenzoate;
$R_1 = p$-HyBz, $R = $ OH
(211) Lancerotol veratrate; $R_1 = $ Ver, $R = $ H
(212) Lancerodiol *p*-methoxybenzoate; $R_1 = p$-Anis,
$R = $ H
(213) Lancerodiol *p*-hydroxybenzoate; $R_1 = p$-HyBz,
$R = $ H
(214) Lancerodiol; $R_1 = R = $ H
(215) Lancerodiol vannillate; $R_1 = $ Van, $R = $ H
(216) Lancerodiol *p*-coumarate; $R_1 = p$-Coum,
$R = $ H
(217) Isolancerodiol *t*-cinnamate; $R_1 = t$-Cinn,
$R = $ H

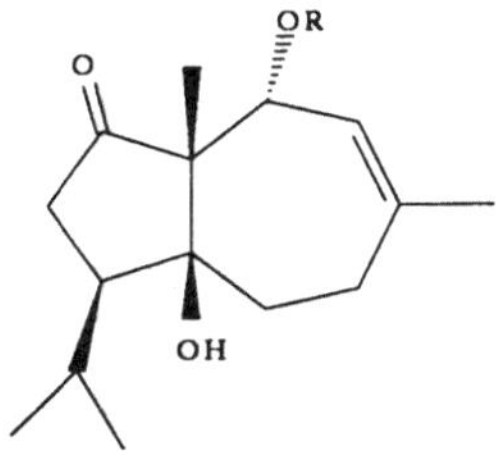

(218) 8-Keto-6-hydroxy-1-*p*-anisoyloxydauc-2-ene
     = (209); R = *p*-Anis
(219) Rel-(3*R*,3a*R*,8*S*,8a*R*)-3a-hydroxy-1-oxo-6-
     daucen-8-yl-*p*-anisate; R = *p*-Anis

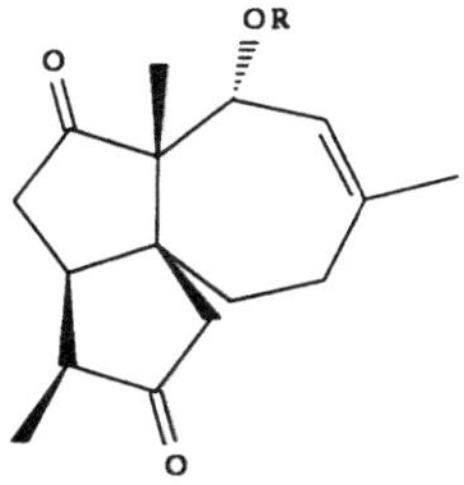

(220) Fercolide; R = *p*-Anis

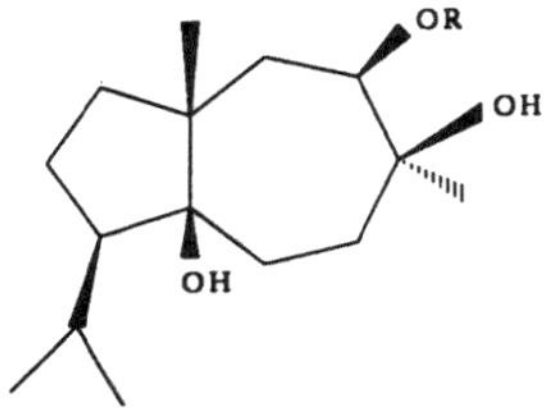

(221) Linkiol; R = Ang
(222) Linkitriol *p*-methoxybenzoate; R = Anis

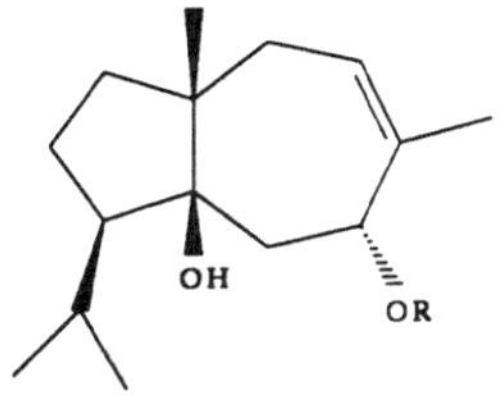

(223) Carotdiol acetate; R = Ac
(224) Carotdiol veratrate; R = Ver

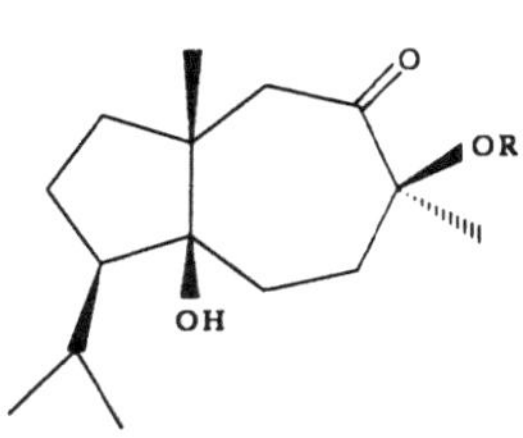

(225) Felikiol 3-angelate; R = Ang
(226) Felikiol; R = H

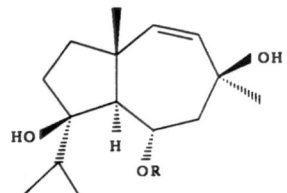

(227)   Webiol angelate; R = Ang
(228)   Webiol epoxyangelate; R = E-Ang

(229)   Ferutriol 5-isovalerate; R = iVal

(230)   Isolancerotetrol 5-isovalerate; R = iVal
(231)   Isolancerotetrol 5-angelate; R = Ang

(232)   Epoxyisolancerotetrol 5-isovalerate; R = iVal

(233)   4β,8β-Dihydroxy-6α-vanilloyloxydauc-9-ene;
        R = Van

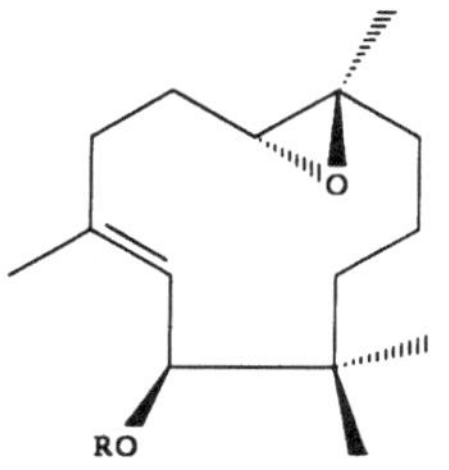

(234) Humulene 1β,10α-epoxy-3-deoxyjuniferol-*p*-anisate; R = *p*-Anis

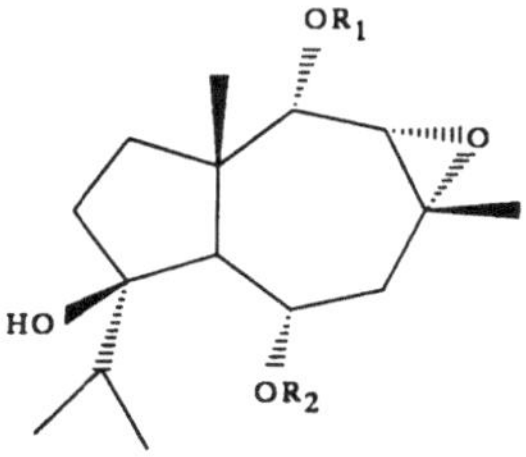

(235) Lapiferol; $R_1 = R_2 = H$
(236) Lapiferin; $R_1 = Ac, R_2 = Ang$

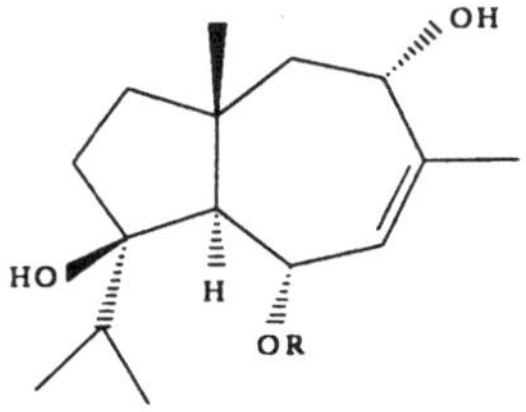

(237) Lancerotriol *p*-hydroxybenzoate; R = *p*-HyBz
(238) Lancerotriol 6-vanillate; R = Van

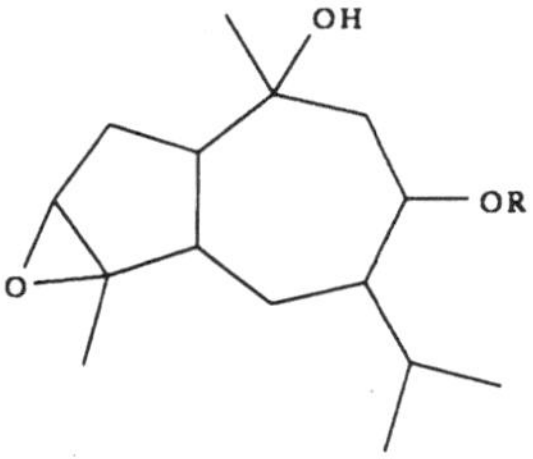

(239) Akiferinin; R = Ver
(240) Tenuferin; R = iVan
(241) Tenufernin; R = Van
(242) Tenuferidin; R = *p*-HyBz

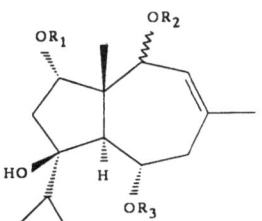

(243) Rel-(1S,3R,3aS,4S,8S,8aR)-1,8-diacetoxy-3-hydroxy-6-daucen-4-yl-*p*-anisate;
$R_1 = R_2 = Ac, R_3 = p$-Anis
(244) Unnamed; $R_1 = Ac, R_2 = β$-Ac, $R_3 = p$-Anis
(245) Unnamed; $R_1 = Ac, R_2 = β$-H, $R_3 = p$-Anis
(246) Unnamed; $R_1 = Ac, R_2 = β$-Ac, $R_3 = H$
(247) Unnamed; $R_1 = Ac, R_2 = α$-Ac, $R_3 = Ver$

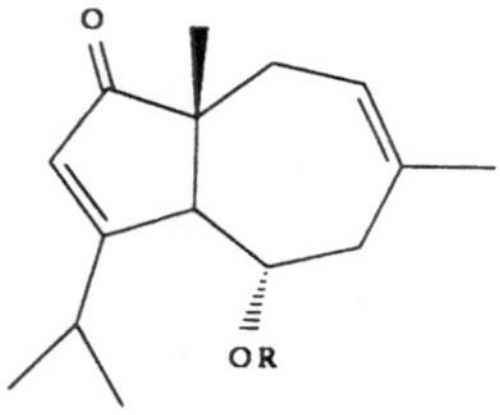

(248) Unnamed; R = *p*-Anis

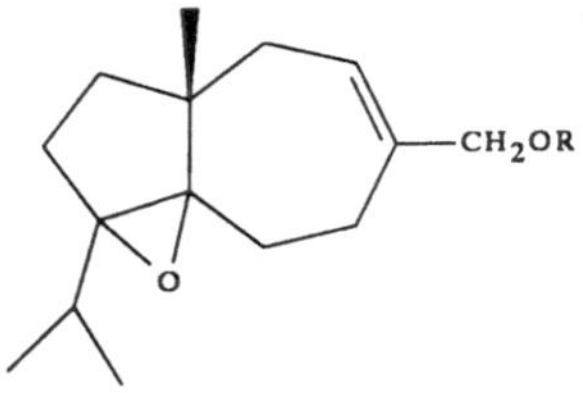

(249) Unnamed; R = *p*-Anis

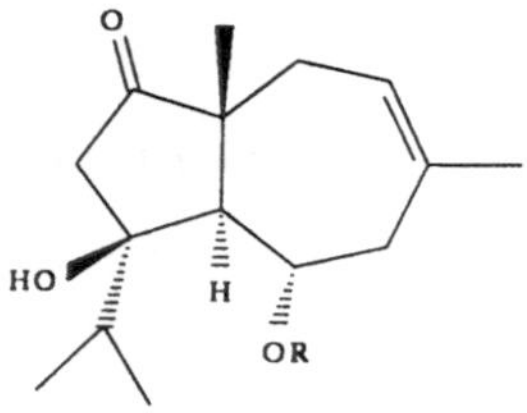

(250) Ferutinone; R = *p*-HyBz

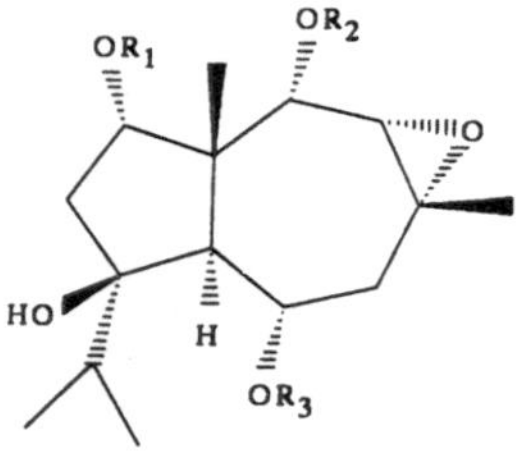

(251) Lapidolinin; $R_1 = R_2 = Ac$, $R_3 = H$
(252) Lapidolin; $R_1 = R_2 = Ac$, $R_3 = Ang$
(253) Lapidolidin; $R_1 = Ac$, $R_2 = H$, $R_3 = Ver$

(254) Lapiferinin; $R_1 = Ac$, $R_2 = Ver$

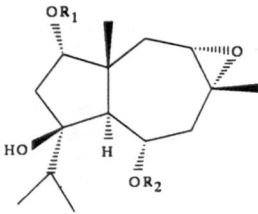

References, pp. 82–92

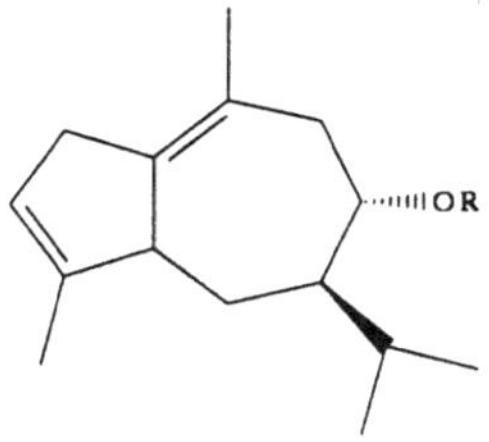

(255) Microferin; R = *p*-HyBz
(256) Microferin; R = Ver

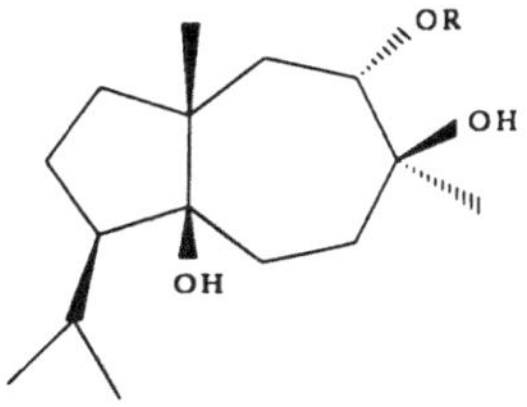

(257) Pallidin; R = Trimethoxygallicate

Undetermined
Structures

(258) Taulin
(259) Taufenin
(260) Taufedin
(261) Tauferin

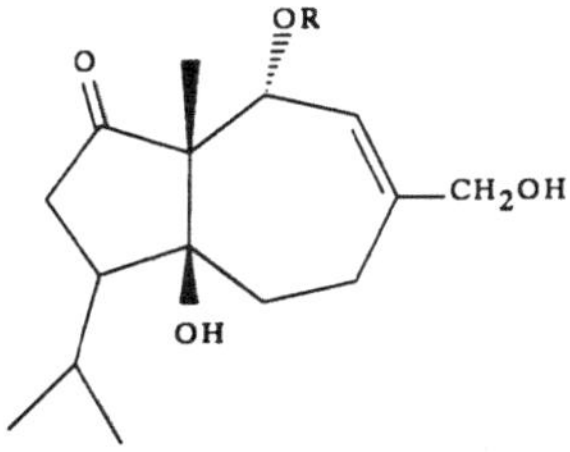

(262) 14-Hydroxyvaginatin R = Ang

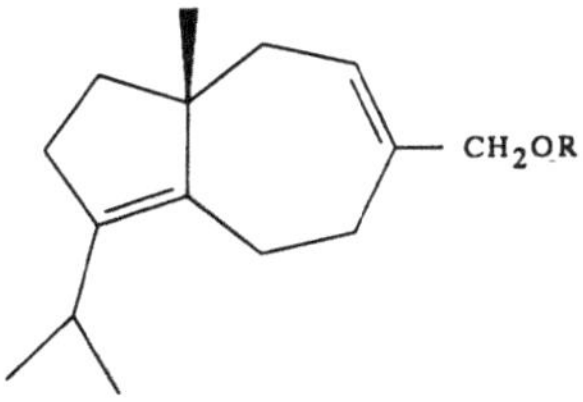

(263) 14-*p*-Anisoyloxydauc-4,8-diene; R = Ans

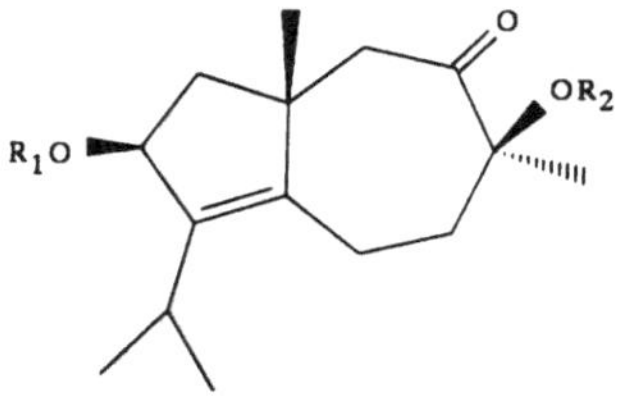

(264) Torosol diangelate; $R_1 = R_2 = Ang$

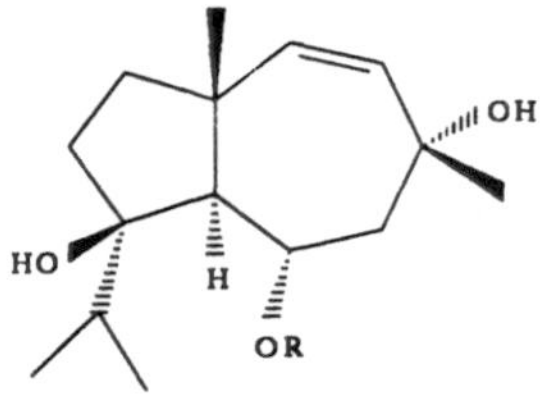

(265) 4β,8α-Dihydroxy-6α-vanilloyloxydauc-9-ene; R = Van

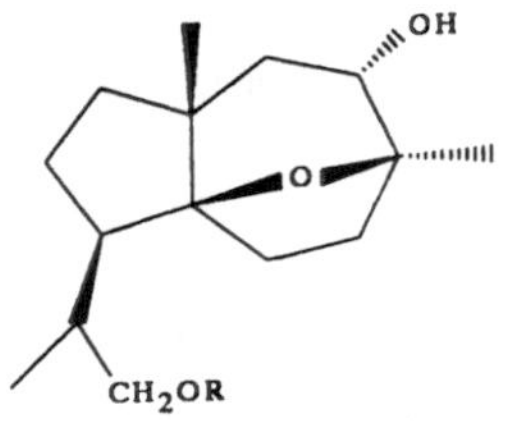

(266) 13-Vanilloyloxydaucol; R = Van

VII. Other Sesquiterpenoids from *Ferula* Species

To round off this article a list of monocyclofarnesane derivatives isolated from *Ferula* species—all ethers of various coumarins—is presented in Table 3 and Chart 20. For a listing of numerous bicyclofarnesyl

Table 3. *Monocyclofarnesane Derivatives Found in Ferula Species*

Structure number	Name of compound	Formula	Plant source	References
(267)	Fekolone	$C_{24}H_{28}O_4$	F. kopetdaghensis	(110)
(268)	Kopetdaghin	$C_{24}H_{30}O_4$	F. kopetdaghensis	(152)
(269)	Fekoline	$C_{26}H_{32}O_4$	F. kopetdaghensis	(151)
(270)	Kopeolin (Fekrol)	$C_{24}H_{32}O_5$	F. kopetdaghensis	(153, 154)
(271)	Feropolone (Kopeolone)	$C_{24}H_{30}O_5$	F. polyantha F. kopetdaghensis	(155, 156) (157)
(272)	Feropolol	$C_{24}H_{34}O_6$	F. polyantha	(155, 156)
(273)	Feropolin	$C_{26}H_{34}O_7$	F. polyantha	(155, 156)
(274)	Feliferin	$C_{24}H_{34}O_6$	F. foliosa	(158)
(275)	Kopeside	$C_{30}H_{42}O_{10}$	F. kopetdaghensis	(153)
(276)	Assafoetidin	$C_{24}H_{30}O_4$	F. assafoetida	(159, 160)

Chart 20. Structure of monocyclofarnesanes found in *Ferula* species

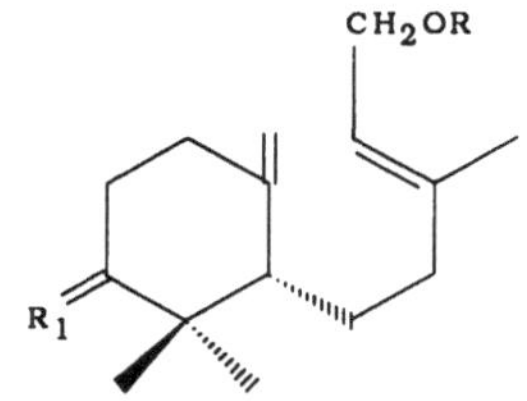

(**267**) Fekolone; R = 7-HyC, R_1 = O

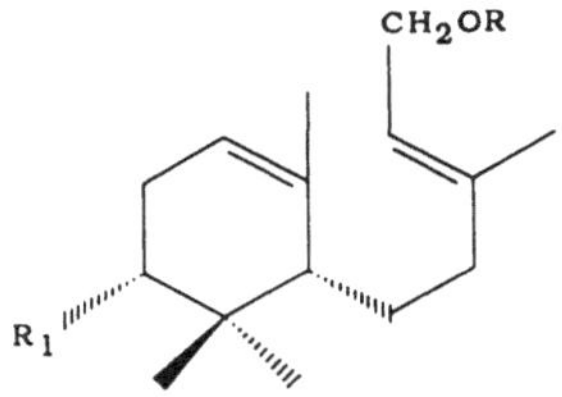

(**268**) Kopetdaghin; R = 7-HyC, R_1 = OH
(**269**) Fekoline; R_1 = OAc, R_2 = 7-HyC

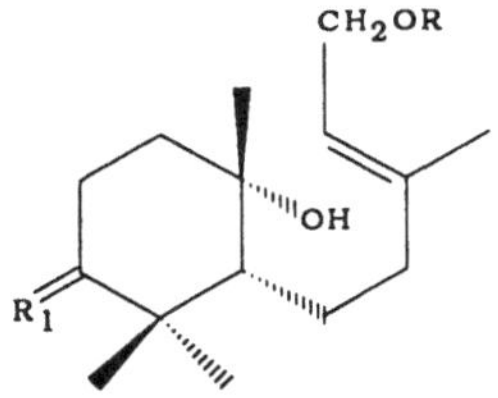

(**270**) Kopeolin; R = 7-HyC, (Fekrol) R_1 = βOH, H
(**271**) Feropolone; R = 7-HyC, (Kopeolone) R_1 = O

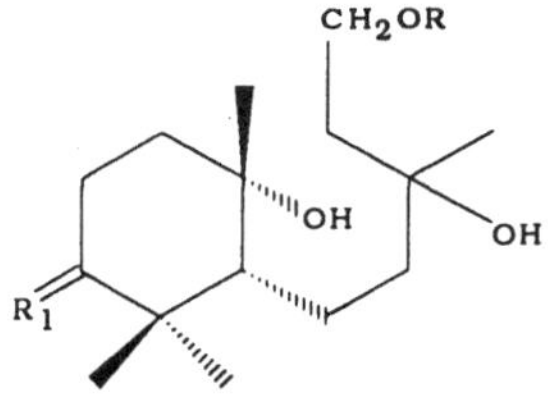

(**272**) Feropolol; R = 7-HyC, R_1 = βOH, H
(**273**) Feropolin; R = 7-HyC, R_1 = βOAc, H
(**274**) Foliferin; R = 7-HyC, R_1 = αOH, H
(**275**) Kopeoside; R = 7-HyC, R_1 = O-β-D-glycoside, H

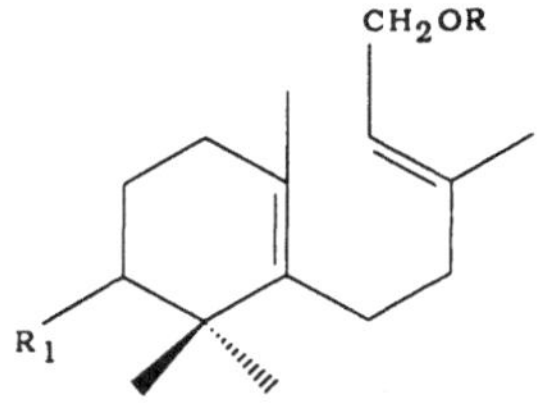

(**276**) Assafoetidin; R = 7-HyC, R_1 = OH

ethers of coumarins including those from *Ferula* species the reviews by MURRAY (*197, 198*) should be consulted. Names and formulas of bicyclic sesquiterpene lactones from *Ferula* species are given in Table 4 and Chart 21.

Table 4. Bicyclic Sesquiterpene Lactones Found in Ferula Species

Structure number	Name of compound	Formula	Plant source	References
(277)	Ferulin	$C_{15}H_{16}O_3$	F. oopoda (Boiss & Buhse) Boiss	(161, 162, 187)
(278)	Ferulidin	$C_{15}H_{18}O_4$	F. oopoda	(163, 187)
(279)	Ferugolide	$C_{28}H_{30}O_8$	F. gigantea B. Fedtsch	(164, 165)
(280)	Giferolide	$C_{27}H_{28}O_7$	F. gigantea	(164, 165)
			F. clematidifolia K. Polj.	(188)
(281)	Gigantolide	$C_{25}H_{26}O_8$	F. gigantea	(164, 165)
(282)	Talassin A	$C_{24}H_{30}O_7$	F. gigantea	(164, 165)
			F. litvinowiana	(166)
			F. malacophylla	(167)
			F. diversivittata	(165, 172, 173)
			F. olgae	(168)
(283)	Fegvolide = (292)	$C_{29}H_{32}O_9$	F. gigantea	(165, 169)
			(F. gigas, F. latifolia)	(189)
(284)	Olgoferin	$C_{23}H_{36}O_7$	F. olgae	(165, 168)
(285)	Olgin	$C_{21}H_{24}O_7$	F. olgae	(165, 168)
(286)	Oferin	$C_{23}H_{28}O_7$	F. olgae	165, 168)
(287)	Laferin	$C_{22}H_{26}O_7$	F. olgae	(165, 168)
			F. varia (Schrenck) Trautv.	(170)
(288)	Talassin B	$C_{23}H_{30}O_7$	F. olgae	(165, 168)
(289)	Malaphyll	$C_{29}H_{32}O_9$	F. malacophylla	(167, 168)
			F. gigantea	(169)
			F. diversivittata	(165)
			F. litvinowiana	(166)
			F. clematidifolia	(188)

(290)	Malaphyllin	$C_{26}H_{28}O_9$	F. malacophylla	(167)
			F. gigantea	(165, 169)
(291)	Malaphyllinin	$C_{24}H_{24}O_7$	F. litvinowiana	(166)
			F. malacophylla	(171)
			F. gigantea	(164, 165)
(292)	Diversolide = (**283**)	$C_{29}H_{32}O_9$	F. diversivittata	(172, 173)
(293)	Grilactone	$C_{15}H_{20}O_2$	F. penninervis Regal & Schmalh.	(174, 175)
			F. oopoda	(175)
			F. grigoriewii B. Fedtsch.	(175, 176)
(294)	Malaphyllidin	$C_{15}H_{12}O_3$	F. malacophylla	(177)
(295)	Ferolide	$C_{22}H_{20}O_4$	F. penninervis	(178, 190)
(296)	Ferugolide	$C_{28}H_{34}O_8$	F. penninervis	(179)
(297)	Jalcaguaianolide deriv.	$C_{15}H_{18}O_2$	F. arrigonii Bocchieri	(180)
(298)	Badkhysin	$C_{20}H_{24}O_5$	F. oopoda	(181)
			F. badkh,si Korov	(165, 191, 192)
(299)	Isobadkhysin	$C_{20}H_{24}O_5$	F. oopoda	(181, 192)
(300)	Shairidin	$C_{20}H_{22}O_5$	F. varia	(165, 170)
(301)	Opoferzin	$C_{20}H_{24}O_5$	F. oopoda	(196)
(302)	Oopodin	$C_{15}H_{20}O_2$	F. oopoda	(182, 183, 193, 194)
(303)	Dehydrooopodin	$C_{20}H_{24}O_4$	F. oopoda	(182, 183, 193)
			F. badkhysi Korov	(184)
(304)	Badkhysidin	$C_{20}H_{26}O_5$	F. oopoda	(182, 183)
(305)	Oxylactone	$C_{15}H_{20}O_4$	F. oopoda	(183, 185)
(306)	Semopodin	$C_{20}H_{24}O_5$	F. oopoda	(183, 186)
(307)	Badkhysinin	$C_{20}H_{24}O_5$	F. oopoda	(182, 183, 193, 194)
(308)	Feropodin	$C_{15}H_{20}O_2$	F. oopoda	(182, 183, 193)
(309)	Opoferdin	$C_{20}H_{24}O_5$	F. oopoda	(195)

Chart 21. Structure of bicyclic sesquiterpenes found in *Ferula* species

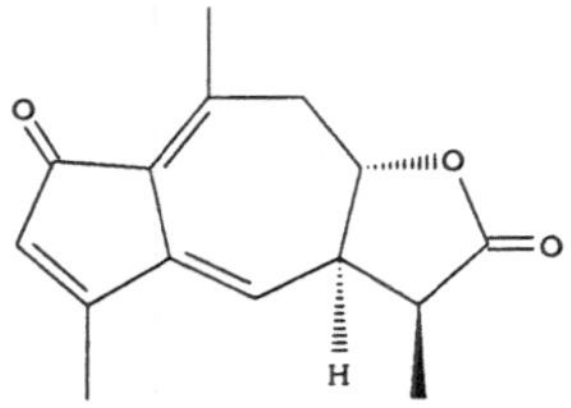

(277) Ferulin

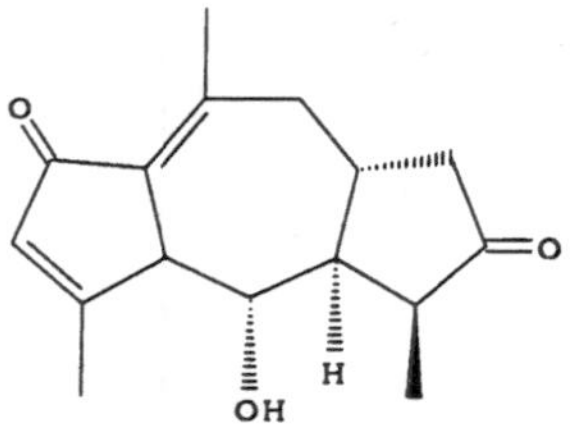

(278) Ferulidin

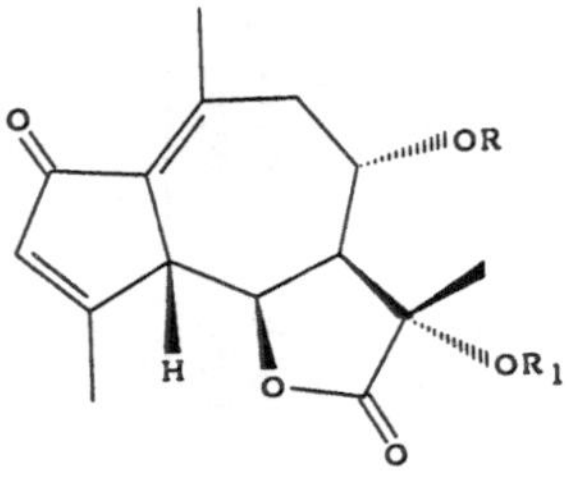

(279) Ferugolide; R = p-Anis, R_1 = Sen
(280) Giferolide; R = Bz, R_1 = Sen
(281) Gigantolide; R = p-Anis, R_1 = Ac
(282) Talassin A; R_1 = R_2 = Ang
(283) Fegvolide = (292); R = Ver, R_1 = Ang
(284) Olgoferin; R = R_1 = Me Acr
(285) Olgin; R = Me Acr, R_1 = Ac
(286) Oferin; R = Me Acr, R_1 = iBut
(287) Laferin; R = Ang, R_1 = Ac
(288) Talassin B; R = Ang, R_1 = iBut
(289) Malaphyll; R = Ver, R_1 = Sen
(290) Malaphyllin; R = Ver, R_1 = Ac
(291) Malaphyllinin; R = Bz, R_1 = Ac

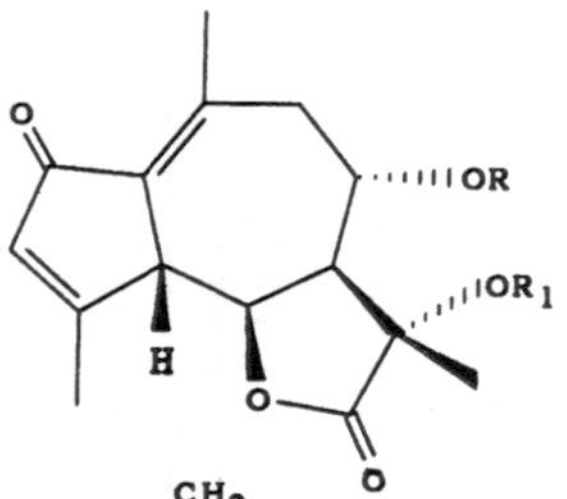

(292) Diversolide = (283); R = Ver, R_1 = Ang

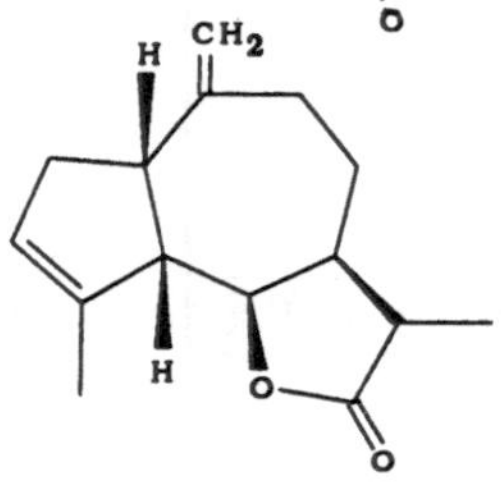

(293) Grilactone

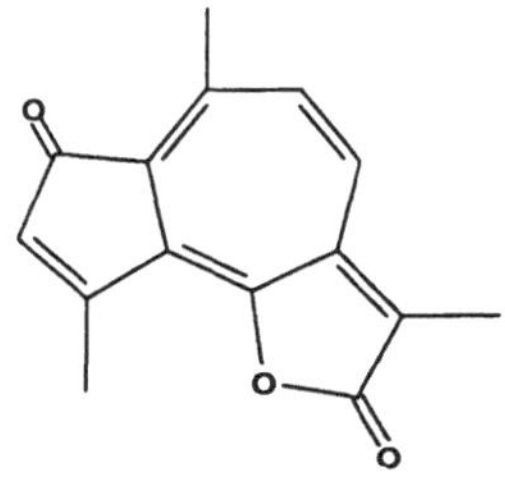

(294) Malaphyllidin

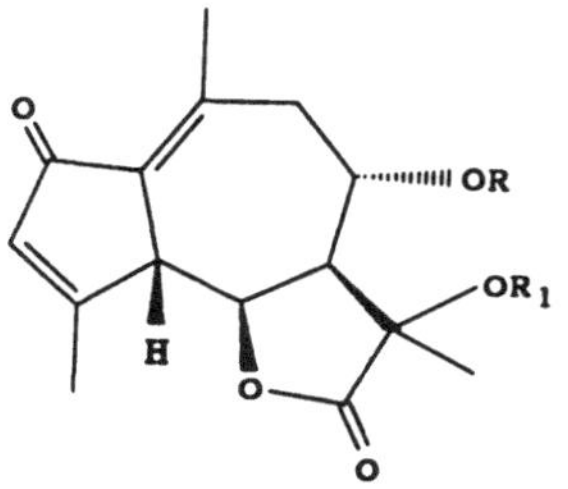

(295) Ferolide; R = Sen, R₁ = Ac

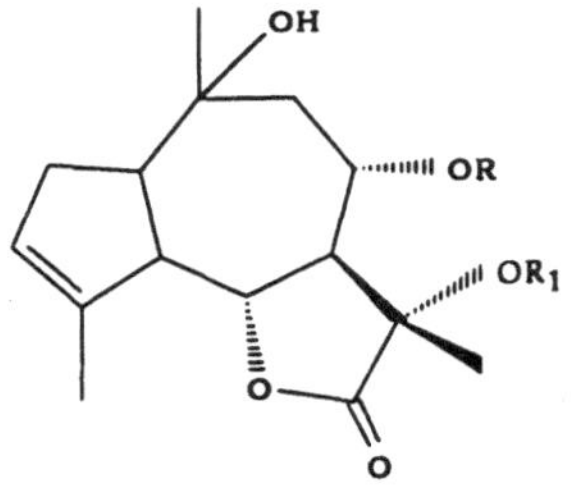

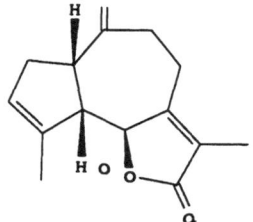

(296) Ferugolide; R = *p*-Anis, R₁ = Sen

(297) Jalcaguaianolide deriv.

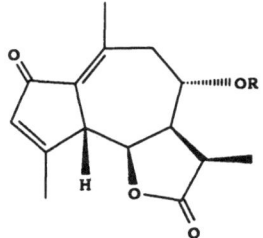

(298) Badkhysin; R = Ang

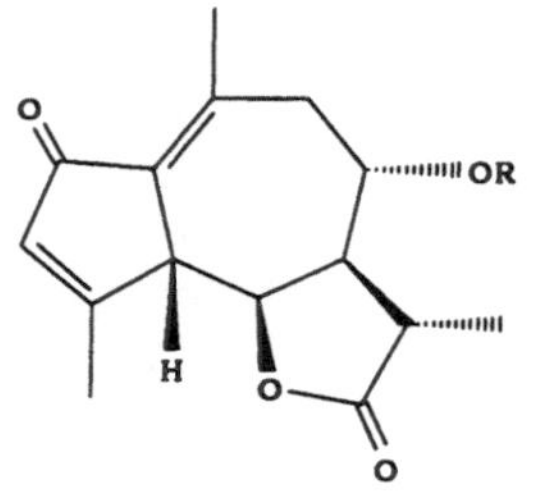

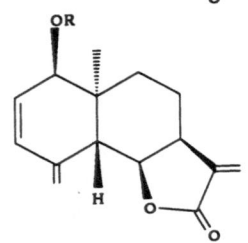

(299)   Isobadkhysin; R = Ang

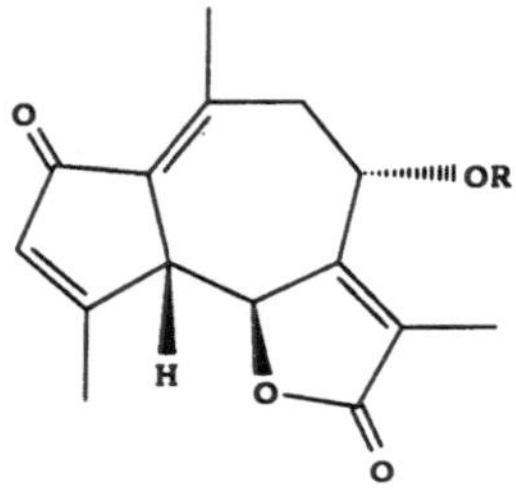

(300)   Shairidin; R = Ang

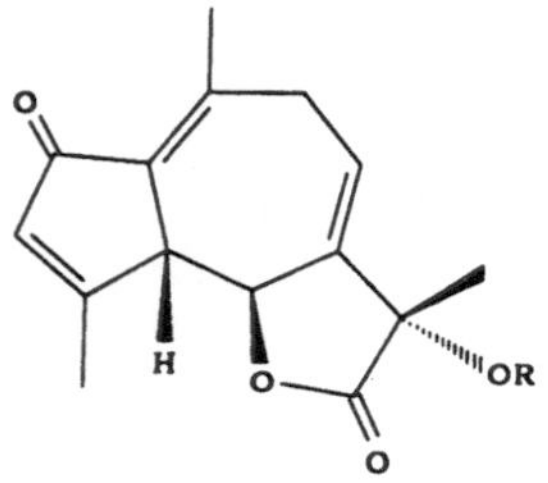

(301)   Opoferzin; R = Ang

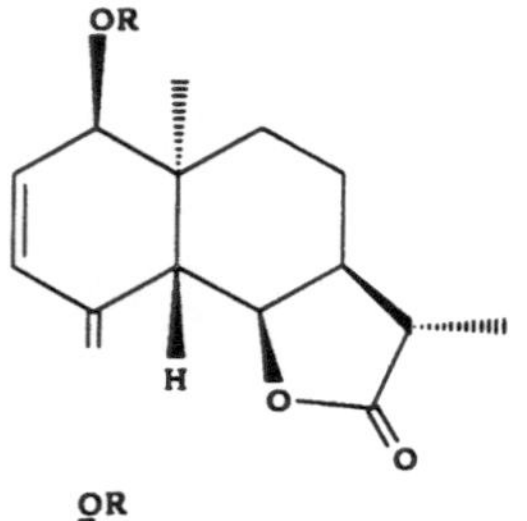

(302)   Oopodin; R = Ang

(303)   Dehydrooopodin; R = Ang

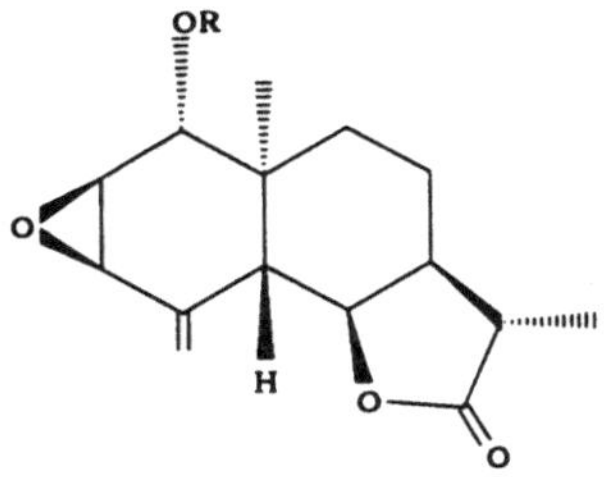

(**304**) Badkhysidin; R = Ang
(**305**) Oxylactone; R = H

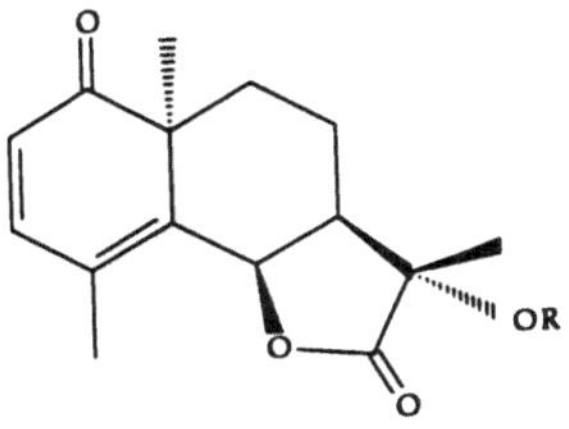

(**306**) Semopodin; R = Ang

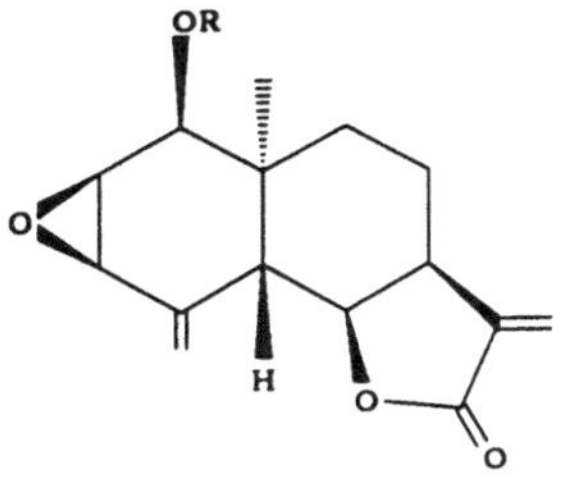

(**307**) Badkhysinin; R = Ang

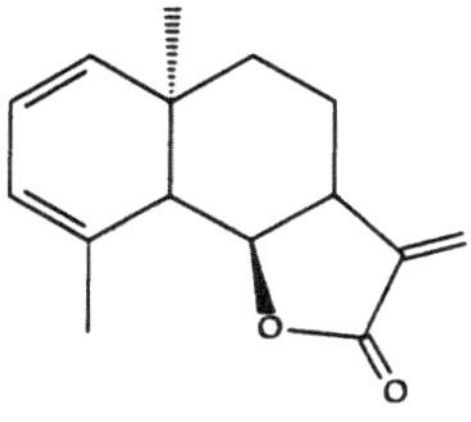

(**308**) Feropodin

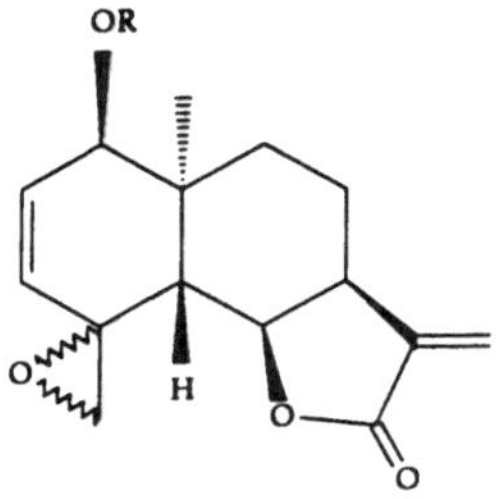

(**309**) Opoferdin; R = Ang

VIII. Common Ester Side Chains in Sesquiterpenes from *Ferula* Species

Angelates or other esters frequently occur in the side chains of *Ferula* sesquiterpenes. To save space we have used the abbreviations in Table 5 throughout.

Table 5. *Common Side Chains in Ferula Sesquiterpenes*

Structure of side chain	Type of ester	Abbreviation
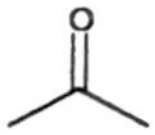	acetate	Ac
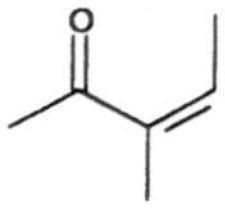	angelate	Ang
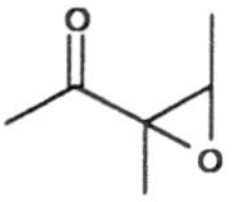	epoxyangelate	E-Ang
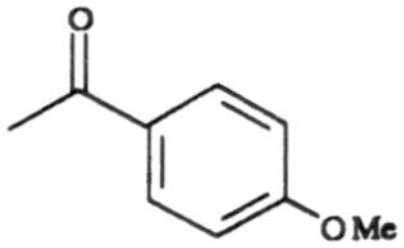	*p*-anisate	*p*-Anis
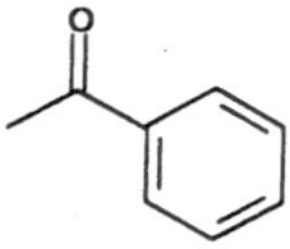	benzoate	Bz
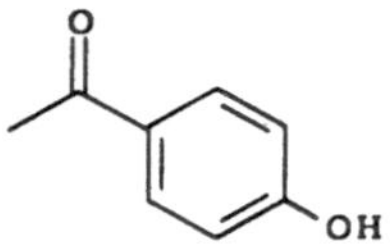	*p*-hydroxybenzoate	*p*-HyBz
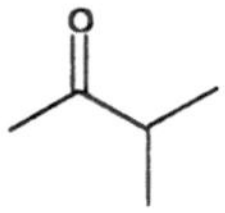	isobutyrate	iBut

Table 5 (*continued*)

Structure of side chain	Type of ester	Abbreviation
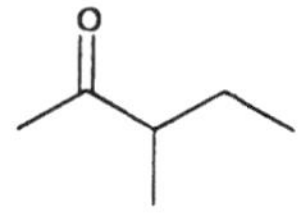	2-methylbutyrate	2-MeBut
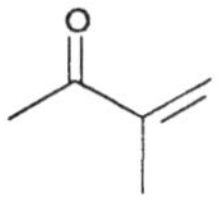	methacrylate	MeAcr
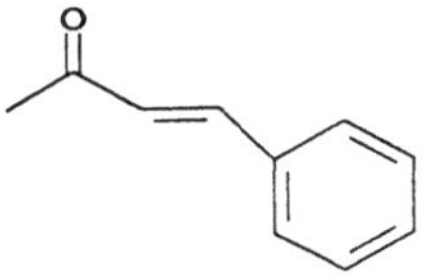	*t*-cinnamate	*t*-Cinn
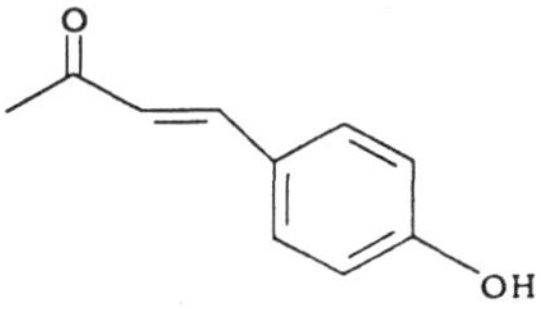	coumarate	Coum
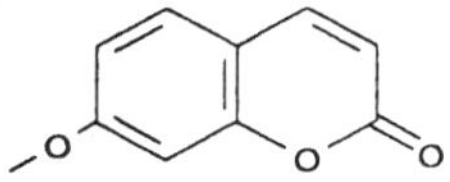	7-hydroxycoumarin	7-HyC
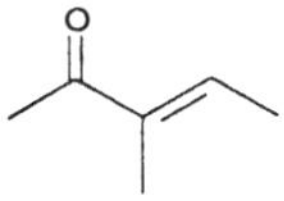	tiglate	Tig
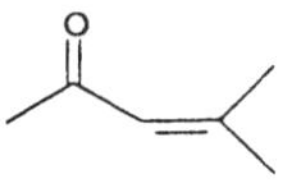	senecioate	Sen
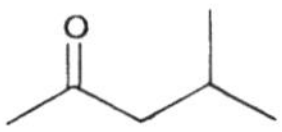	isovalerate	iVal

Table 5 (*continued*)

Structure of side chain	Type of ester	Abbreviation
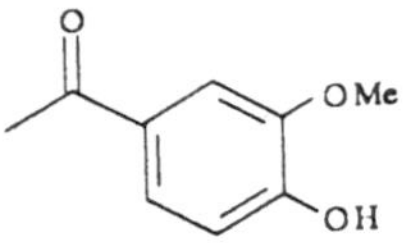	vanillate	Van
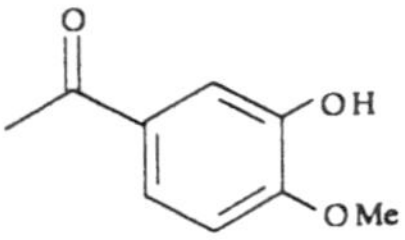	isovanillate	iVan
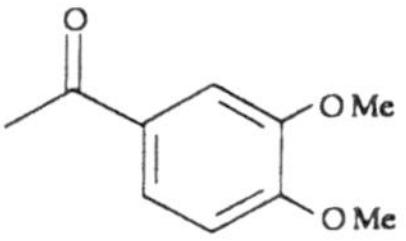	veratrate	Ver

Acknowledgements

This work has been partly financed by a grant from CICYT (PB 91-0148) while AGG is indebted to the Colegio Libre de Eméritos (Madrid). We should also like to thank Professor Werner Herz who has shown a more than merely editorial interest in this manuscript.

References

1. Korovin, E.P.: Illustrated Monograph of the Genus *Ferula*. Tashkent: 1947. See ref. 28.
2. Saidkhodzhaev, A.I.: Sesquiterpene Derivatives of the Genus *Ferula*. Khim. Prir. Soedin., **4**, 437 (1979); Chem. Abstr., **92**, 76670p (1980).
3. Kir'yalov, N.P.: Coumarins from Plants of the Genus *Ferula*. Tr. Botan. Inst., Akad. Nauk SSSR, Ser. 5, Rast. Syr'e, **12**, 82 (1965); Chem. Abstr., **59**, 5255h (1966).
4. Kir'yalov, N.P., and V.Yu. Bagirov: Structure of Karatavikin. Khim. Prir. Soedin., **3**, 223 (1967); Chem. Abstr., **68**, 68829m (1968).
5. Kir'yalov, N.P., and V.Yu. Bagirov: Structure of Karatavic Acid. Khim. Prir. Soedin., **4**, 283 (1968); Chem. Abstr., **70**, 77715q (1969).
6. Kir'yalov, N.P., and S.D. Movchan: Structure of Samarkandine and Samarkandone, Coumarin Compounds of *Ferula samarkandica*. Khim. Prir. Soedin., **4**, 73 (1968); Chem. Abstr., **69**, 59040r (1968).
7. Kir'yalov, N.P.: The Coumarin, Badrakemin, from the Roots of *Ferula badrakema*. Khim. Prir. Soedin., **3**, 363 (1967); Chem. Abstr., **69**, 35865n (1968).

8. KIR'YALOV, N.P., and S.D. MOVCHAN: Structure of Gummosine. Khim. Prir. Soedin., **2**, 383 (1966); Chem. Abstr., **67**, 11394r (1967).

9. KIR'YALOV, N.P., and S.V. SERKEROV: Structure of Badkhysin. Khim. Prir. Soedin., **4**, 341 (1968); Chem. Abstr., **70**, 78177c (1969).

10. KIR'YALOV, N.P.: Species of the Genus *Ferula* as Sources of New Biologically Active Compounds. Tr. Bot. Inst., Akad. Nauk. SSSR, Ser. 5, **15**, 129 (1968); Chem. Abstr., **27**, 3185r (1969).

11. PIMENOV, M.G., YU.E. SKLYAR, A.A. SAVINA, and YU.V. BARANOVA: Chemosystematics of Section Palaeonarthex of the Genus *Ferula*. Biochem. Syst. Ecol., **10**, 133 (1982).

12. ASHRAF, M., R. AHMAD, S. MAHAMOOD, and M.K. BHATTY: Studies on the Essential Oils of the Pakistani Species of the Family Umbelliferae, XLV: *Ferula assafoetida, Linn* (Herra Hing) Gum Oil. Pak. J. Sci. Ind. Res., **23**, 68 (1980); Chem. Abstr., **94**, 127135u (1981).

13. COURCHET, L.: Les Ombellifères en Général et les Espèces Usitées en Pharmacie en Particulier. Montpellier: Imprimerie Cristin, Serre and Ricome. 1982.

14. UPHOF, J.C.TH.: Dictionary of Economic Plants, 2nd Ed. Verlag von J. Cramer. 1968.

15. CULBRETH, D.M.R.: A Manual of Materia Medica and Pharmacology, 6th Ed. Philadelphia and New York: Lea and Febiger. 1917.

16. AL-HAZINI, H.M.G.: Sesquiterpenoids of *Ferula* Species. J. Chem. Soc. Pak., **10**, 482 (1988)

17. HAAGEN-SMITH, A.J.: The Essential Oils. New York: Van Nostrand. 1948.

18. RUZICKA, L.: The Isoprene Rule and Biogenesis of Terpenic Compounds. Experientia, **9**, 357 (1953).

19. HENDRICKSON, J.B.: Stereochemical Implications in Sesquiterpene Biogenesis. Tetrahedron, **7**, 82 (1959).

20. AGVANOFF, B.W., H. EGGEREN, U. HENNING, and F. LYNEN: Isopentenol Pyrophosphate Isomerase. J. Amer. Chem. Soc., **81**, 1254 (1959).

21. WOODWARD, R.B.: Structure and the Absorption Spectra of α,β-Unsaturated Ketones. J. Amer. Chem. Soc., **63**, 1123 (1941).

22. GOLOVINA, L.A., and A.I. SAIDKHODZHAEV: Esters of *Ferula rubroarenosa*. Khim. Prir. Soedin., **6**, 712 (1978); Chem. Abstr., **91**, 2498w (1979).

23. MISKI, M., T.J. MABRY, and O. SAYA: New Daucane and Germacrane Esters from *Ferula orientalis* var. *orientalis*. J. Nat. Prod., **50**, 829 (1987).

24. KARAVAEVA, O.V., V.V. KISELEVA, G.K. NIKONOV, and M.O. KARRYEV: Esters of *Ferula ovina* (Boiss). Izv. Akad. Nauk Kaz. SSR, Ser. Khim., **1**, 80 (1989); Chem. Abstr., **111**, 4208g (1989).

25. CORDELL, G.A.: Biosynthesis of Sesquiterpenes. Chem. Rev., **76**, 425 (1976).

26. SOUCEK, M.: On Terpenes, CXLVIII: Biosynthesis of Carotol in *Daucus carota* L. A Contribution to Configuration of Carotol and Daucol. Collect. Czech. Chem. Commun., **27**, 2929 (1962).

27. SAGITDINOVA, G.V., A.I. SAIDKHODZHAEV, and V.M. MALIKOV: Structure of Fertinidine and Fertinicine. Khim. Prir. Soedin., **1**, 42 (1980); Chem. Abstr., **93**, 46865j (1980).

28. KOROVIN, E.: Genesis *Ferula* (Tourn) L. Monographia Illustrata, pp. 9, 59. Tashkent: Editio Academia Scientarum USSR. 1947.

29. MISKI, M., T.J. MABRY, and O. SAYA: Apiene Esters from *Ferula haussknechtii*. Phytochem., **26**, 1733 (1987).

30. BOHLMANN, F., C. ZDERO, and M. GRENZ: Neue Sesquiterpene der *Gattung Othonna*. Chem. Ber., **107**, 3928 (1974).

31. BOHLMANN, F., C. ZDERO, R.M. KING, and H. ROBINSON: Humulene Derivatives from *Acritoppapus prunifolius*. Phytochem., **21**, 147 (1982).
32. EL DAHMY, S., J. JAKUPOVIC, F. BOHLMANN, and T.M. SARG: New Humulene Derivatives from *Asteriscus graveolens*. Tetrahedron, **41**, 309 (1985).
33. GONZÁLEZ, A.G., J. BERMEJO, J.G. DIAZ, and L. ARANCIBIA: Humulenes and Other Constituents of *Ferula latipinna*. J. Nat. Prod., **51**, 1140 (1988).
34. ITOKAWA, H., H. MATSUMOTO, S. MISHASHI, and Y. IITAKA: Two Novel Humulenoids from *Torilis scabra D.C.*. Chem. Lett., 1581 (1983).
35. ALAN, F.T., M. OZAINNE, R. DECORZANT, and F. NÄF: 10-Epijunenol, a New Cis Eudesmane Sesquiterpenoid. Tetrahedron, **32**, 2261 (1976).
36. SENDA, I., and S. IMAIZUMI: ^{13}C Pulse Fourier Transform NMR of Menthol Stereoisomers and Related Compounds. Tetrahedron, **31**, 2905 (1975).
37. FELTKAMP, H., and N.C. FRANKLIN: Conformational Analysis by Nuclear Magnetic Resonance, VI: The Conformations of Sixteen Stereomeric 2,5-Dialkyl Cyclohexanols of the Menthol Type. Tetrahedron, **21**, 1541 (1965).
38. BROWN, H.C., C.P. GARG, and K.T. LIU: The Oxidation of Secondary Alcohols in Diethyl Ether with Aqueous Chromic Acid. A Convenient Procedure for the Preparation of Ketones in High Epimeric Purity. J. Org. Chem., **36**, 387 (1971).
39. ASAKAWA, Y., G. OURISSON, and T. ARATANI: New Sesquiterpene Lactone and Aldehyde of *Frullania tamarisci* subsp. *oscura* (Hepatieae). Tetrahedron Letters, **45**, 3957 (1975).
40. NIWA, M., A. NISHIYAMA, M. IGUCHI, and S. YAMAMURA: Sesquiterpenes from *Acorus calamus*. Bull. Chem. Soc. Japan, **48**, 2930 (1975).
41. NÄF, F., R. DECORZANT, and W. THOMMEN: Regiospecific Intramolecular Aldol Condensation Induced by Conjugate Addition of Lithium Dimethylcuprate to C-oxo-α,β-Enones. Helv. Chim. Acta, **58**, 1808 (1975).
42. BIZHANOVA, K., and A.I. SAIDKHODZHAEV: The Structure of Xeroferin. Khim. Prir. Soedin., **5**, 581 (1978); Chem. Abstr., **90**, 104150y (1979).
43. SRIRAMANAN, M.C., B.A. NAGASAMPAGI, R.C. PANDEY, and S. DEV: Studies in Sesquiterpenes, XLIX: Sesquiterpenes from *Ferula jaeschkeana* Vatke (Part I): Jaeschkeanadiol-Structure, Stereochemistry. Tetrahedron, **29**, 985 (1973).
44. CHIURDOGLU, G., and M. DESCAMPS: Contribution a l'Etude des Composes Sesquiterpeniques, IV: Etude de la Structure du Carotol, Alcool $C_{15}H_{26}O$ de l'Essence de *Daucus carota*. Tetrahedron, **8**, 271 (1960).
45. HOLUB, M., J. TAX, P. SEDMERA, and F. SORM: On Terpenes, CCVII: Contribution to the Stereochemistry of Substances with Daucane Skeleton and the Determination of Absolute Configuration of Laserpitine. Coll. Czech. Chem. Comm., **35**, 3597 (1970).
46. BUDZIKIEWICZ, H., C. DJERASSI, and D.N. WILLIAMS: Mass Spectrometry of Organic Compounds, p. 94. San Francisco: Holden-Day. 1967.
47. CURTIS, R.G., I. HEILBRON, E.R.H. JONES, and G.F. WOODS: The Chemistry of Triterpenes, XIII: The Further Characterization of Polyporenic Acid A. J. Chem. Soc., 457 (1953).
48. HOLUB, M., Z. SAMEK, V. HEROUT, and F. SORM: On Terpenes, CLXXXIII: The Constitution of Laserpitine. A Sesquiterpenic Compound from *Laserpitium latifolium* L. Root. Coll. Czech. Chem. Comm., **32**, 591 (1967).
49. TAKEDA, K., and H. MINATO: Sesquiterpenoids, I: Absolute Configuration of Guaiol. Chem. Pharm. Bull., **9**, 619 (1961); Chem. Abstr., **56**, 10193d (1962).
50. BHACCA, N.S., and D.H. WILLIAMS: Applications of NMR Spectroscopy in Organic Chemistry, pp. 82–83. San Francisco: Holden-Day. 1964.
51. GARG, S.N., S.K. AGARWAL, V.P. MAHAJAN, and S.N. RASTOGI: Feruginidin and

Ferugin, Two New Sesquiterpenoids Based on the Carotane Skeleton from *Ferula jaeschkeana*. J. Nat. Prod., **50**, 253 (1987).

52. GARG, S.N., and S.K. AGARWAL: Further New Sesquiterpenes from *Ferula jaeschkeana*. J. Nat. Prod., **51**, 771 (1988).
53. GARG, S.N., VISHWAPAUL, and S.N. RASTOGI: Isocarotane and Carotane Derivatives from *Ferula jaeschkeana*. Phytochem., **29**, 531 (1990).
54. MISKI, M., A. ULUBELEN, and T.J. MABRY: Six Sesquiterpene Alcohol Esters from *Ferula elaeochytris*. Phytochem., **22**, 2231 (1983).
55. DIAZ, J.G., B.M. FRAGA, A.G. GONZÁLEZ, P. GONZÁLEZ, and M.G. HERNANDEZ: Eight Carotane Sesquiterpenes from *Ferula linkii*. Phytochem., **23**, 2541 (1984).
56. DIAZ, J.G., B.M. FRAGA, A.G. GONZÁLEZ, M.G. HERNANDEZ, and A. PERALES: Carotane Sesquiterpenes from *Ferula linkii*. Phytochem., **25**, 1161 (1986).
57. GARG, S.N., and K. AGARWAL: New Sesquiterpenes from *Ferula jaeschkeana*. Planta Med., **53**, 341 (1987).
58. BIZHANOVA, K.B., A.I. SAIDKHODZAEV, and V.M. MALIKOV: Structure of Jaeschferin. Khim. Prir. Soedin., **7**, 127 (1980); Chem. Abstr., **93**, 71992f (1980).
59. AHMED, A.A.: New Sesquiterpenes from *Ferula sinaica*. J. Nat. Prod., **53**, 483 (1990).
60. MISKI, M., A. ULUBELEN, T.J. MABRY, W.H. WATSON, I. VICKOVIC, and M. HOLUB: A New Sesquiterpene Ester from *Ferula tingitana*. Tetrahedron, **40**, 5197 (1984).
61. HOLUB, M., Z. SAMEK, V. HEROUT, and F. SORM: Terpenes, CLXXXVII: Constitution of Isolaserpitin, Deoxodehydrolaserpitin and Laserpitinol. Monatsh. Chemie, **98**, 1138 (1967).
62. MISKI, M., and T.J. MABRY: New Daucane Esters from *Ferula tingitana*. J. Nat. Prod., **49**, 657 (1986).
63. MISKI, M., and T.J. MABRY: Daucane Esters from *Ferula communis* subsp. *communis*. Phytochem., **24**, 1735 (1985).
64. MISKI, M., and T.J. MABRY: Fercolide, A Type of Sesquiterpene Lactone from *Ferula communis* subsp. *communis* and the Correct Structure of Vaginatin. Phytochem., **25**, 1673 (1986).
65. HOLUB, M., R. DE GROOTE, V. HEROUT, and F. SORM: Plant Substances, XXVIII: Oxygen-Containing Components of Light Petroleum Extract of *Laser trilobum* (L.) Borkh Root. Structure of Laserine. Coll. Czech. Chem. Comm., **33**, 2911 (1968).
66. PINAR, M., M. RICO, and B. RODRIGUEZ: Laserine Oxide, an Epoxide from *Guillonea scraba*. Phytochem., **21**, 735 (1982).
67. FRAGA, B.M., A.G. GONZÁLEZ, P. GONZÁLEZ, M.G. HERNANDEZ, and C.L. LARRUGA: Carotane Sesquiterpenes from *Ferula lancerottensis*. Phytochem., **24**, 501 (1985).
68. HASSAN, A., and G. AL-HAZIMI: More Sesquiterpenes from *Ferula sinaica*. Oriental. J. Chem., **4**, 200 (1988).
69. AHMED, A.A.: Daucanes and Other Constituents from *Ferula sinaica*. Phytochem., **30**, 1207 (1991).
70. FRAGA, B.M., M.G. HERNANDEZ, J.G. DIAZ, A.G. GONZÁLEZ, and P. GONZÁLEZ: Carotane Sesquiterpenes from *Ferula linkii*. Phytochem., **25**, 2883 (1986).
71. LAMNAOUER, D., M. MARTIN, D. MALHO, and B. BODO: Isolation of Daucane Esters from *Ferula communis* var. *berivifolia*. Phytochem., **28**, 2711 (1989).
72. RAZDAN, T.K., B. QADRI, M.A. QURISHI, M.A. KHUROO, and P.K. KACHROO: Sesquiterpene Esters and Sesquiterpene-Coumarin Ethers from *Ferula jaeschkeana*. Phytochem., **28**, 3389 (1989).
73. ZALKOW, L.H., M.G. CLOWER, M.M. GORDON, and L.T. GELBAUM: Chemistry of the Carotane Sesquiterpenes, V: Biosimulated Conversions. J. Nat. Prod., **43**, 382 (1980).
74. ZALKOW, L.H., M.G. CLOWER JR., M. GORDON, J. SMITH, D. VAN DER VEER, and J. A.

BERTRAND: Formic Acid-Catalysed Rearrangement of Carotol. X-Ray Crystal Structure of the *p*-Iodobenzoate of One of the Alcohols Formed. J. Chem. Soc. Chem. Commun., 374 (1976).

75. YAMASAKI, M.: Total Synthesis of the Sesquiterpene (−)-Daucene. J. Chem. Soc. Commun., 606 (1972).

76. DE BROISSIA, H., J. LEVISALLES, and H. RUDLER: Total Synthesis of the Sesquiterpenes (+)-Daucene, (+)-Carotol, and (−)-Daucol. J. Chem. Soc. Chem. Commun., 855 (1972).

77. MAHMOUD, Z.F., N.A. ABDEL SALAM, T.M. SARG, and F. BOHLMANN: A Carotane Derivative and a Eudesmanolide from *Inula crithmoides*. Phytochem., 20, 735 (1981).

78. ULUBELEN, A., N. GÖREN, F. BOHLMANN, J. JAKUPOVIC, M. GRENZ, and N. TANKER: Sesquiterpene Lactones from *Smyrnium cordifolium*. Phytochem., 24, 1305 (1985).

79. GONZÁLEZ, A.G., J.G. DIAZ, L.A. LOPEZ, P.P. DE PAZ, E. VALENCIA, and J. BERMEJO: Sesquiterpene Alcohol Esters and Sesquiterpene-Coumarin Ethers from *Ferula linkii-TF* (to be published).

80. GONZÁLEZ, A.G., J.G. DIAZ, L.A. LOPEZ, P.P. DE PAZ, and J. BERMEJO: Distribution of Secondary Metabolites in *Ferula* from the Canary Islands (to be published).

81. SANTOS, A.: Vegetación y flora de La Palma. Santa Cruz de Tenerife: Int. Canar. 1983.

82. P. DE PAZ, P.: Personal communication.

83. GONZÁLEZ, A.G., B.M. FRAGA, M.G. HERNANDEZ, J.G. LUIS, R. ESTEVEZ, J.L. BAEZ, and M. RIVERO: Linkiol, a New Sesquiterpene from *Ferula linkii*. Phytochem., 16, 265 (1977).

84. DE BROISSIA, H., J. LEVISALLES, and H. RUDLER: Terpénoides-V, Carotol-IV. Synthèse du (+) Daucène, du (+) Carotol et du (−) Daucol. Bull. Soc. Chim. France, 11, 4314 (1972).

85. FRAGA, B.M., M.G. HERNANDEZ, and J.G. DIAZ: Carotdiol Esters from *Ferula linkii*. Phytochem., 28, 1649 (1989).

86. LEVISALLES, J., and H. RUDLER: Terpénoïdes-II, Carotol-I: Stéréochimie du Carotol. Bull. Soc. Chim. France, 6, 2059 (1967).

87. APPENDINO, G., M.G. VALLE, and P. GARIBOLDI: Structural and Conformational Studies on Sesquiterpenoid Esters from *Laserpitium halleri* Crantz subsp. *halleri*. J. Chem. Soc. Perkin Trans. I, 1363 (1986).

88. SAIDKHODZHAEV, A.I., and G.R. NIKONOV: Esters in *Ferula kuhistanica* Roots. Khim. Prir. Soedin., 4, 525 (1974); Chem. Abstr., 82, 54180v (1975).

89. SYKORA, V., L. NOVOTNY, M. HOLUB, V. HEROUT, and F. SORM: Terpenes, CXVII: The Proof of Structure of Carotol and Daucol. Coll. Czech. Chem. Commun., 26, 788 (1961); Chem. Abstr., 55, 16588a (1962).

90. COREY, E.J., and G. SCHMIDT: Useful Procedures for the Oxidation of Alcohols Involving Pyridinium Dichromate in Aprotic Media. Tetrahedron Letters, 399 (1979).

91. SUNDARARAMAN, P., and HERZ, W.: Oxidative Rearrangements of Tertiary and Some Secondary Allylic Alcohols with Chromium (VI) Reagents. A New Method for 1,3 Functional Group Transposition and Forming Mixed Aldol Products. J. Org. Chem., 42, 813 (1977).

92. GOLOVINA, L.A., and A.I. SAIDKHODZHAEV: Structure and Stereochemistry of Lapidin. Khim. Prir. Soedin., 3, 318 (1981); Chem. Abstr., 95, 204188p (1981).

93. GONZÁLEZ, A.G.: Personal communication.

94. SAGITDINOVA, G.V., A.I. SAIDKHODZHAEV, V.M. MALIKOV, and S. MELIBAEV: Esters from *Ferula angrenii*. Khim. Prir. Soedin., 6, 808 (1978); Chem. Abstr., 91, 16666m (1979).

95. KERIMOV, S.SH., A.I. SAIDKHODZHAEV, and V.M. MALIKOV: Esters of *Ferula calcarea*. Khim. Prir. Soedin., 5, 765 (1987); Chem. Abstr., 108, 91725c (1988).

96. SAIDKHODZHAEV, A.I., KH.M. KAMILOV, and M.G. PIMENOV: Esters of *Ferula kopetdaghensis*. Khim. Prir. Soedin., **5**, 764 (1987); Chem. Abstr., **108**, 91724b (1988).

97. SAGITDINOVA, G.V., A.I. SAIDKHODZHAEV, G.K. NIKONOV, and M. RAKHMANKULOV: Esters from *Ferula lapidosa* Roots. Khim. Prir. Soedin., **11**, 115 (1975); Chem. Abstr., **83**, 111086r (1975).

98. KHASANOV, T.KH., A.I. SAIDKHODZHAEV, and G.K. NIKONOV: Components of *Ferula pallida* Roots. Khim. Prir. Soedin., **6**, 807 (1972); Chem. Abstr., **78**, 94809f (1973).

99. POTAPOV, V.M., and G.K. NIKONOV: Chemical Study of *Ferula tenuisecta Eug.* Izv. Akad. Nauk Kaz. SSR, Ser. Khim., 69 (1980); Chem. Abstr., **93**, 110550g (1980).

100. KADYROV, A.SH., T.KH. KHASANOV, A.I. SAIDKHODZAEV, and G.K. NIKONOV: New Phenol Compounds of *Ferula tschimganica* Roots. Khim. Prir. Soedin., **6**, 808 (1972); Chem. Abstr., **78**, 108207x (1973).

101. KAMILOV, KH.M., G.K. NIKONOV, and N.N. SHARAKHIMOV: Esters in *Ferula fedtschenkoana* Roots. Khim. Prir. Soedin., **4**, 527 (1974); Chem. Abstr., **82**, 14012k (1975).

102. SAIDKHODZHAEV, A.I., L.A. GOLOVINA, V.M. MALIKOV, S. MALIBAEV, and M. RAKHMANKULOV: Esters from Three *Ferula* Species. Khim. Prir. Soedin., **3**, 412 (1985); Chem. Abstr., **103**, 51224d (1985).

103. GOLOVINA, L.A., and G.K. NIKONOV: Esters of *Ferula ceratophylla*. Khim. Prir. Soedin., **5**, 707 (1977); Chem. Abstr., **88**, 85988t (1978).

104. GOLOVINA, L.A., A.I. SAIDKHODZHAEV, V.M. MALIKOV, and M.G. PIMENOV: Esters of *Ferula soonganica* and *Ferula subtilis*. Khim. Prir. Soedin., **5**, 763 (1987); Chem. Abstr., **108**, 91723a (1988).

105. SAGITDINOVA, G.V., A.I. SAIDKHODZHAEV, V.M. MALIKOV: Components of *Ferula tschatcalensis* Roots. Khim. Prir. Soedin., **6**, 721 (1983); Chem. Abstr., **100**, 171546v (1984).

106. MAKHMUDOV, M.K., B. TASHKHODZHAEV, G.V. SAGITDINOVA, and A.I. SAIDKHOD-ZHAEV: X-Ray Analysis of Sesquiterpene Esters from Plants of the Genus *Ferula*, II: Crystalline and Molecular Structures of the Humulan Alcohol Fekserol. Khim. Prir. Soedin., **1**, 42 (1986); Chem. Abstr., **105**, 24111z (1986).

107. SAIDKHODZHAEV, A.I., G.V. SAGITDINOVA, L.A. GOLOVINA, and A.SH. KADYROV: Structure of Novel Esters from the *Ferula* Genus of Plants. Tezisy Dokl. Sov. Indiiskii Simp. Khim. Prir. Soedin., **5**, 80 (1978); Chem. Abstr., **93**, 146293q (1980).

108. SAGITDINOVA, G.V., A.I. SAIDKHODZHAEV, and G.K. NIKONOV: Phenolic Components of *Ferula juniperina*. Khim. Prir. Soedin., **4**, 547 (1976); Chem. Abstr., **85**, 189198q (1976).

109. SAIDKHODZHAEV, A.I., N.D. ABDULLAEV, T.KH. KHASANOV, G.K. NIKONOV, and M.R. YAGUDAEV: Structure of Angrendiol, Ferolin, and Tschimganidin. Khim. Prir. Soedin., **4**, 519 (1977); Chem. Abstr., **88**, 3049a (1978).

110. KADYROV, A.SH., A.I. SAIDKHODZHAEV, and G.K. NIKONOV: Structure of Fecorine, a New Ester from *Ferula korshinskyi* Roots. Khim. Prir. Soedin., **11**(2), 152 (1975); Chem. Abstr., **83**, 111106x (1975).

111. POTAPOV, V.M., KH.M. KAMILOV, and G.K. NIKONOV: Phytochemistry Study of *Ferula involucrata*. Nek. Probl. Farm. Nauki Prakt., Mater. S'ezda Farm. Kas., **1**, 112 (1975); Chem. Abstr., **87**, 2357s (1977).

112. POTAPOV, V.M., and G.K. NIKONOV: Esters of *Ferula leucographa* and *Ferula ugamica*. Izv. Akad. Nauk Kaz. SSR, Ser. Khim., 64 (1981); Chem. Abstr., **95**, 21286z (1981).

113. BAGIROV, V.YU., R.YU. GASANOVA, F.A. RASULOV, and M.G. PIMENOV: Ester from *Ferula leucographa*. Khim. Prir. Soedin., **3**, 422 (1980); Chem. Abstr., **93**, 164328j (1980).

114. SAGITDINOVA, G.V., A.I. SAIDKHODZHAEV, and V.M. MALIKOV: Structure of Fertenin. Khim. Prir. Soedin., **2**, 161 (1979); Chem. Abstr., **92**, 6723j (1980).

115. POTAPOV, V.M., and G.K. NIKONOV: Esters of *Ferula tenuisecta.* Eug. Kor. from the Aksu-Dzhabagli Wildlife Refuge. Izv. Akad. Nauk Kaz. SSR, Ser. Khim., **2**, 63 (1980); Chem. Abstr., **93**, 110549p (1980).

116. BIZHANOVA, K., A.I. SAIDKHODZHAEV, and V.M. MALIKOV: Fecseridin and Fecserinin-—New Esters from *Ferula xeromorpha.* Khim. Prir. Soedin., **5**, 576 (1978); Chem. Abstr., **90**, 83600c (1979).

117. BIZHANOVA, K., A.I. SAIDKHODZHAEV, and V.M. MALIKOV: Structure of Fekserine. Khim. Prir. Soedin., **3**, 407 (1978); Chem. Abstr., **89**, 147091b (1978).

118. KADYROV, A., A.I. SAIDKHODZHAEV, and G.K. NIKONOV: Structure of Akitschenol and Akitschenin. Khim. Prir. Soedin., **1**, 102 (1976); Chem. Abstr., **85**, 59565w (1976).

119. SAIDKHODZHAEV, A.I., D. BATSUREN, and V.M. MALIKOV: Esters from *Ferula akitschensis.* Khim. Prir. Soedin., **5**, 709 (1985); Chem. Abstr., **104**, 85436k (1986).

120. SAIDKHODZHAEV, A.I., and G.K. NIKONOV: Components of *Ferula ovina* Roots. Khim. Prir. Soedin., **4**, 526 (1974); Chem. Abstr., **82**, 14011j (1975).

121. POTAPOV, V.M., and G.K. NIKONOV: Esters of *Ferula pallida* Roots. Khim. Prir. Soedin., **6**, 819 (1976); Chem. Abstr., **86**, 136306n (1977).

122. AHMED, A.A.: New Sesquiterpenes from *Ferula sinaica.* J. Nat. Prod., **53**, 483 (1990).

123. EIDLER, YA.I., G.L. GENIKINA, and T.T. SHAKIROV: Chromatospectrophotomeric Method for Determination of Ferutinin, Teferin and Ferutin. Khim. Prir. Soedin., **5**, 616 (1976); Chem. Abstr., **86**, 85646p (1977).

124. MELIBAEV, S., U. RAKHMANKULOV, and A.I. SAIDKHODZHAEV: Intraspecies Chemical Variability in *Ferula tenuisecta.* Korov. Rastit. Resur., 431 (1980); Chem. Abstr., **93**, 182807u (1980).

125. KAZDAN, T.K., B. QADRI, M.A. QURISHI, M.A. KHUROO, and P.K. KACHROO: Sesquiterpene Esters and Sesquiterpene-Coumarin Ethers from *Ferula jaeschkeana.* Phytochem., **28**, 3389 (1989).

126. GARG, S.N., L.N. MISRA, S.K. AGARWAL, V.P. MAHAJAN, and S.N. RASTOGI: Carotane Derivatives from *Ferula jaeschkeana.* Phytochem., **25**, 449 (1987).

127. SINGH, M.M., A. AGNIHOTRI, S.N. GARG, S.K. AGARWAL, D.N. GUPTA, G. KESHRI, and V.P. KAMBOJ: Antifertility and Hormonal Properties of Certain Carotane Sesquiterpenes of *Ferula jaeschkeana.* Planta Med., **54**(6), 492 (1988).

128. KADYROV, A.SH., A.I. SAIDKHODZHAEV, G.K. NIKONOV, and U.R. KULOV: Esters of *Ferula akitschensis.* Khim. Prir. Soedin., **2**, 284 (1977); Chem. Abstr., **87**, 98862b (1977).

129. KADYROV, A.SH., and A.I. SAIDKHODZHAEV: Structure of Akiferinin. Khim. Prir. Soedin., **1**, 137 (1978); Chem. Abstr., **88**, 191130s (1978).

130. SAIDKHODZHAEV, A.I.: Structure of Tenuferin, Tenuferinin and Tenuferidin. Khim. Prir. Soedin., **1**, 70 (1978); Chem. Abstr., **89**, 6437g (1978).

131. WATSON, W.H., R.P. KASHYAP, and I. TAVANAIEPOUR: rel-(3R,3aR,8S,8aR)-3a-hydroxyl-1-oxo-6-daucen-8-yl *p*-anisate(1) and rel-(1S,3R,3aS,4S,8S,8aR)-1,8-diacetoxy-3-hydroxy-6-daucen-4-yl *p*-anisate(2). Acta Crystallogr., Sect. C: Cryst. Struct. Commun., 1050 (1985); Chem. Abstr., **104**, 6038j (1986).

132. VALLE, M.G., A. GIOVANNI, N.M. GIAN, and P. VICENZO: Prenylated Coumarins and Sesquiterpenoids from *Ferula communis.* Phytochem., **26**, 253 (1987).

133. SAIDKHODZHAEV, A.I., and G. NIKONOV: Structure of Teferidine, a New Ester from *Ferula tenuisecta* Fruit. Khim. Prir. Soedin., **1**, 105 (1976); Chem. Abstr., **85**, 59567y (1976).

134. MISKI, M., and J. JAKUPOVIC: Daucane Esters from *Ferula rigidula.* Phytochem., **29**, 173 (1990).

135. SAIDKHODZHAEV, A.I., and G.K. NIKONOV: Esters in *Ferula kuhistanica* Roots. Khim. Prir. Soedin., **4**, 525 (1974); Chem. Abstr., **82**, 54180v (1975).

136. KHASANOV, T.KH., A.I. SAIDKHODZHAEV, and G.K. NIKONOV: Structure of Teferin, a New Ester from *Ferula tenuisecta* Root. Khim. Prir. Soedin., **4**, 528 (1974); Chem. Abstr., **82**, 82954u (1975).

137. GOLOVINA, L.A., A.I. SAIDKHODZHAEV, V.M. MALIKOV, and S. MELIBAEV: Structure of Lapidolin and Lapidolidin. Khim. Prir. Soedin., **6**, 787 (1982); Chem. Abstr., **98**, 143673f (1983).

138. TASHKHODZHAEV, B., M.K. MAKHMUDOV, L.A. GOLOVINA, A.I. SAIDKHODZHAEV, and M.R. YAGUDAEV: X-Ray Structure Characteristics of Sesquiterpene Esters from *Ferula* Plants, I: Crystal and Molecular Structure of the Lapidoline Carotane Ester. Khim. Prir. Soedin., **3**, 309 (1984); Chem. Abstr., **102**, 59336s (1985).

139. GOLOVINA, L.A., A.I. SAIDKHODZHAEV, and V.M. MALIKOV: Structure and Stereochemistry of Lapiferinin. Khim. Prir. Soedin., **3**, 301 (1983); Chem. Abstr., **99**, 102267d (1983).

140. GOLOVINA, L.A., A.I. SAIDKHODZHAEV, and V.M. MALIKOV: Structure of Lapidolidin. Khim. Prir. Soedin., **5**, 713 (1985); Chem. Abstr., **104**, 126521e (1986).

141. GOLOVINA, L.A., A.I. SAIDKHODZHAEV, N.D. ABDULLAEV, V.M. MALIKOV, and M.R. YAGUDAEV: Structure and Stereochemistry of Lapiferin. Khim. Prir. Soedin., **3**, 296 (1983); Chem. Abstr., **99**, 102266c (1983).

142. GOLOVINA, L.A., T.KH. KHASANOV, A.I. SAIDKHODZHAEV, V.M. MALIKOV, and U. RAKHMANKULOV: Coumarins and Esters from *Ferula microcarpa*. Khim. Prir. Soedin., **5**, 566 (1978); Chem. Abstr., **90**, 83598h (1979).

143. KUSHMURADOV, A.YU., M.K. MAKHMUDOV, A.I. SAIDKHODZHAEV, B. TASHKHOD-ZHAEV, V.M. MALIKOV, and M.R. YAGUDAEV: X-Ray Structural Analysis of Sesquiterpene Esters from Plants of the Genus *Ferula*, IV: Structure and Stereochemistry of the New Carotane Ester Pallidin. Khim. Prir. Soedin., **1**, 42 (1990); Chem. Abstr., **113**, 55865t (1990).

144. KUSHMURADOV, A.YU., A.I. SAIDKHODZHAEV, and A.SH. KADYROV: Structure of Palliferidin from *Ferula pallida*. Khim. Prir. Soedin., **3**, 400 (1981); Chem. Abstr., **95**, 204189q (1981).

145. KUSHMURADOV, A.YU., A.I. SAIDKHODZHAEV, and A.SH. KADYROV: Structure and Stereochemistry of Palliferidin and Palliferinin. Khim. Prir. Soedin., **4**, 523 (1981); Chem. Abstr., **96**, 52525b (1982).

146. KUSHMURADOV, A.YU., A.I. SAIDKHODZHAEV, and V.M. MALIKOV: Structure and Stereochemistry of Pallinin. Khim. Prir. Soedin., **1**, 53 (1986); Chem. Abstr., **105**, 75874h (1986).

147. AL-HAZIMI, H.M.G.: Terpenoids and Coumarins from *Ferula sinaica*. Phytochem., **10**, 2417 (1986).

148. MAKHMUDOV, M.K., L.A. GOLOVINA, B. TASHKODZHAEV, A.I. SAIDKHODZHAEV, M.R. YAGUDAEV, and V.M. MALIKOV: X-Ray Structure Investigation of Sesquiterpene Esters from Plants of the Genus *Ferula*. Khim. Prir. Soedin., **1**, 68 (1989); Chem. Abstr., **111**, 74798y (1989).

149. KUSHMURADOV, A.YU., A.SH. KADYROV, A.I. SAIDKHODZHAEV, and V.M. MALIKOV: Structure of Akiferidin and Akiferidinin. Khim. Prir. Soedin., **6**, 725 (1978); Chem. Abstr., **90**, 200272k (1979).

150. RAJENDRA, K., S.K. PAKNIKAR, G.K. TRIVEDI, and S.C. BHATTACHARYYA: Structure of Vaginatin, A Polyoxygenated Sesquiterpenoid Isolated from *Selinum vaginatum*. Ind. J. Chem., **16B**, 4 (1978).

151. NABIEV, A.A., T.KH. KHASANOV, and V.M. MALIKOV: New Terpenoid Coumarins from *Ferula kopetdaghensis*. Khim. Prir. Soedin., **4**, 516 (1978); Chem. Abstr., **89**, 193845y (1978).

152. KAMILOV, KH.M., and G.K. NIKONOV: Coumarins of *Ferula kopetdaghensis*, Kopetdaghin and Farnesiferol B (Copeodin). Khim. Prir. Soedin., **4**, 442 (1974); Chem. Abstr., **82**, 70266h (1974).

153. KAMILOV, KH.M., and G.K. NIKONOV: Coumarins of *Ferula kopetdaghensis* and Structure of Kopedine and Kopeoside. Khim. Prir. Soedin., **3**, 308 (1973); Chem. Abstr., **79**, 134321v (1973).

154. VESELOVSKAYA, N.V., YU.E. SKLYAR, D.A. FESENKO, and M.G. PIMENOV: Fekrol, A Terpenoid Coumarin from *Ferula krylovii*. Khim. Prir. Soedin., **6**, 851 (1979); Chem. Abstr., **93**, 41500b (1980).

155. KHASANOV, T.KH., A.I. SAIDKHODZHAEV, and G.K. NIKONOV: Structure of Feropolin, Feropolol, Feropolon and Feropolidin. Khim. Prir. Soedin., **1**, 91 (1976); Chem. Abstr., **85**, 59558w (1976).

156. SAIDKHODZHAEV, A.I., and V.M. MALIKOV: Stereochemistry of Feropolol, Feropolin, Feropolone and Feropolidin. Khim. Prir. Soedin., **6**, 799 (1978); Chem. Abstr., **90**, 200277r (1979).

157. NABIEV, A.A., T.KH. KHASANOV, and V.M. MALIKOV: New Terpenoid Coumarin from *Ferula kopetdaghensis*. Khim. Prir. Soedin., **1**, 48 (1982); Chem. Abstr., **96**, 214275z (1982).

158. KADYROV, A.SH., A.I. SAIDKHODZHAEV, and V.M. MALIKOV: Structure of Foliferin. Khim. Prir. Soedin., **4**, 518 (1978); Chem. Abstr., **90**, 69072a (1979).

159. BANERJI, A., B. MALLICK, and A. CHATTERJEE: Assafoetidin and Ferocolicin, Two Sesquiterpenoid Coumarins from *Ferula assafoetida* Regel. Tetrahedron Letters, **29**, 1557 (1988).

160. KAJIMOTO, T., K. YAHIRO, and T. NOHARA: Sesquiterpenoid and Disulphide Derivatives from *Ferula assafoetida*. Phytochem., **28**, 1761 (1989).

161. SERKEROV, S.V.: Structure of Ferulin. Khim. Prir. Soedin., **6**, 371 (1970); Chem. Abstr., **73**, 88041e (1970).

162. SERKEROV, S.V.: Ferulin, Sesquiterpene Lactone from *Ferula oopoda* Roots. Khim. Prir. Soedin., **6**, 134 (1970); Chem. Abstr., **73**, 73111d (1970).

163. SERKEROV, S.V.: Structure of Ferulidin, A New Sesquiterpenoid Lactone. Khim. Prir. Soedin., **6**, 428 (1970); Chem. Abstr., **74**, 17861g (1971).

164. SAVINA, A.A., D.A. FESENKO, L.I. DUKHOVLINOVA, YU.E. SKLYAR, M.G. PIMENOV, and YU.V. BARANOVA: Lactones of *Ferula gigantea*. Khim. Prir. Soedin., **4**, 490 (1979); Chem. Abstr., **92**, 111177g (1980).

165. RYCHLEWSKA, U., D.J. HODSON, M. HOLUB, M. BUDESINSKY, and Z. SMITALOVA: The Structure of 2-Oxo-8α-angeloyloxy-11α-acetoxy-5βH,6αH,7αH-guai-1(10),3-dien-6,12-olide, A Sesquiterpenic Lactone from *Laserpitium prutenicum* L. Revision of the Stereostructures of Native 2-Oxoguai-1(10),3-dien-6,12-olides from the Species of the Umbelliferae Family. Coll. Czech. Comm., **50**, 2607 (1985).

166. BAGIROV, V.YU., V.I. SHEICHENKO, N.F. MIR-BABAEV, and M. PIMENOV: Sesquiterpene Lactones of *Ferula litvinowiana*. Khim. Prir. Soedin., **1**, 114 (1984); Chem. Abstr., **100**, 188795y (1984).

167. BAGIROV, V.YU., V.I. SHEICHENKO, R.YU. GASANOVA, and M.G. PIMENOV: Study of *Ferula malacophylla* Lactones. Khim. Prir. Soedin., **4**, 445 (1978); Chem. Abstr., **89**, 211929k (1978).

168. KONOVALOVA, O.A., K.S. RYBALKO, and V.I. SHEICHENKO: Sesquiterpene Lactones of *Ferula olgae*. Khim. Prir. Soedin., **5**, 590 (1975); Chem. Abstr., **84**, 105802a (1976).

169. SAVINA, A.A., L.I. DUKHOVLINOVA, YU.E. SKLYAR, D.A. FESENKO, and M.G. PIMENOV: Fegnolide A Lactone from *Ferula gigantea* Roots. Khim. Prir. Soedin., **5**, 733 (1979); Chem. Abstr. **94**, 61703u (1981).

170. BAGIROV, V.YU., V.I. SHEICHENKO, I.K. ABDULLAEVA, and M.G. PIMENOV: Sesquiterpene Lactones of *Ferula varia*. Khim. Prir. Soedin., **6**, 843 (1980); Chem. Abstr., **94**,
171054r (1981).
171. BAGIROV, V.YU., V.I. SHEICHENKO, R.YU. GASANOVA, and M.G. PIMENOV: Sesquiterpene Lactone from *Ferula malacophylla* Seeds. Khim. Prir. Soedin., **6**, 810 (1978);
Chem. Abstr., **91**, 2502t (1979).
172. KISELEVA, V.V., A.I. SAIDKHODZHAEV, and G.K. NIKONOV: Sesquiterpene Lactones
from the Roots of *Ferula diversivittata*. Khim. Prir. Soedin., **1**, 126 (1975); Chem.
Abstr., **84**, 5166c (1976).
173. SERKEROV, S.V., and V.V. KISELEVA: Structure of Diversolide. Khim. Prir. Soedin., **4**,
525 (1982); Chem. Abstr., **98**, 125719w (1983).
174. NURMUKHAMEDOVA, R.M., SH.Z. KASYMOV, and S. MALIBAEV: Grilactone from
Ferula penninervis. Khim. Prir. Soedin., **2**, 261 (1982); Chem. Abstr., **97**, 107075h (1982).
175. MALONE, F.J., M. PARVES, A. KARIN, A.M. MCKERVEY, I. AHMAD, and K.M. BHATTY:
Isolation and Crystal Structure of Grilactone, A New Guaianolide from *Ferula
oopoda*. J. Chem. Soc. Perkin Trans. 2, **11**, 1683 (1980).
176. KIR'YALOV, N.P., and S.V. SERKEROV: Guaianolides from Plants of the Umbelliferae
Family. Mezhdumar. Kongr. Efirnym. Maslam. [Mater] (pub. 1971), **1**, 147 (1968);
Chem. Abstr., **78**, 159896d (1973).
177. BAGIROV, V.YU., V.I. SHEICHENKO, R.YU. GASANOVA, and M.G. PIMENOV: Sesquiterpene Lactone from *Ferula malacophylla*. Khim. Prir. Soedin., **6**, 811 (1978); Chem.
Abstr., **91**, 2503u (1979).
178. NURMUKHAMEDOVA, M.R., SH.Z. KASYMOV, and G.P. SIDYAKIN: Ferolide, A New
Lactone from *Ferula penninervis*. Khim. Prir. Soedin., **4**, 533 (1983); Chem. Abstr., **100**,
3491e (1984).
179. NURMUKHAMEDOVA, M.R., SH.Z. KASYMOV, N.D. ABDULLAEV, and G.P. SIDYAKIN:
Structure of Fegurolide. Khim. Prir. Soedin., **3**, 335 (1985); Chem. Abstr., **103**, 85070r
(1985).
180. CASINOVI, C.G., L. TOMASSINI, and M. NICOLETTI: A New Guaianolide from *Ferula
arrigonii* Bocchieri. Gass. Chim. Ital., **11**, 563 (1989); Chem. Abstr., **112**, 175650v (1990).
181. SERKEROV, S.V.: Stereochemistry of Guaianolide from *Ferula oopoda*. Khim. Prir.
Soedin., **5**, 629 (1980); Chem. Abstr., **94**, 117786t (1981).
182. SERKEROV, S.V.: Stereochemistry of *Ferula oopoda* Eudesmanolides. Khim. Prir.
Soedin., **4**, 510 (1980); Chem. Abstr., **94**, 65869a (1981).
183. HOLUB, M., M. BUDESINSKY, Z. SMITLOVA, D. SAMAN, and U. RYCHLEWSKA: Structure
of Isosilerolide, Relative and Absolute Configuration of Silerolide and Larolide
Sesquiterpenic Lactones of New Stereoisomeric Type of Eudesmanolides. Coll. Czech.
Chem. Comm., **51**, 903 (1986).
184. SERKEROV, S.V.: Lactone Sesquiterpene of *Ferula badkhysi*. Khim. Prir. Soedin., **3**, 393
(1976); Chem. Abstr., **85**, 139718p (1976).
185. SERKEROV, S.V.: New Sesquiterpene Hydroxylactone from *Ferula oopoda*. Khim. Prir.
Soedin., **6**, 838 (1971); Chem. Abstr., **76**, 151049t (1972).
186. SERKEROV, S.V.: Sesquiterpene Lactone, Semopodin, from *Ferula oopoda* Seeds. Khim.
Prir. Soedin., **4**, 241 (1969); Chem. Abstr., **72**, 63626r (1970).
187. SERKEROV, S.V.: Stereochemistry of *Ferula oopoda*. Khim. Prir. Soedin., **16**, 629 (1980);
Chem. Abstr., **94**, 117786t (1981).
188. SAGITDINOVA, G.V., A.I. SAIDKHODZHAEV, V.M. MALIKOV, M.G. PIMENOV, and S.
MELIBAEV: Sesquiterpene Lactones of *Ferula clematidifolia* and *Lipularia alpina*.
Khim. Prir. Soedin., **26**, 553 (1990); Chem. Abstr., **114**, 98274n (1991).
189. SAVINA, A.A., D. OVLINOVA, YU.E. SKLYAR, D.A. FESENKO, and M.G. PIMENOV: A

Lactone from the Roots of *Ferula gigantea*. Khim. Prir. Soedin., **15**, 733 (1980); Chem. Abstr., **94**, 61703u (1981).

190. TASHKHODZHAEV, B., M.K. MAKHMUDOV, M.R. NORMUKHAMEDOVA, and S.A. TALI- POV: Crystal and Molecular Structure of the Sesquiterpene Lactone Ferulide. Khim. Prir. Soedin., **24**, 55 (1988); Chem. Abstr., **109**, 211243p (1988).

191. SERKEROV, S.V.: Lactones of *Ferula badkhysi*. Khim. Prir. Soedin., **12**, 393 (1976); Chem. Abstr. **85**, 139718p (1976).

192. RYCHLEWSKA, U., and S.V. SERKEROV: Structural Characterization of Badkhysin and Its C(5)-Epimer, Isobadkhysin. Acta Cryst., **C47**, 1872 (1991).

193. SERKEROV, S.V.: Stereochemistry of the Eudesmanolides from *Ferula oopoda*. Khim. Prir. Soedin., **16**, 510 (1980); Chem. Abstr., **94**, 65869a (1981).

194. RYCHLEWSKA, U., B. SZCZEPANSKA, and V. SERKEROV: Two Eudesmanolide Type Sesquiterpene Lactones from Umbelliferae. Acta Cryst., **C48**, 1543 (1992).

195. SERKEROV, S.V., A.N. ALESKEROVA, D.M. AKHMEDOV, and A.M. RASULOV: A New Sesquiterpene Lactone-Opoferdin from *Ferula oopoda*. Khim. Prir. Soedin., **28**, 284 (1992).

196. SERKEROV, S.V., U. RYCHLEWSKA, A.N. ALESKEROVA, and N.F. MIR-BABAEV: A New Guaianolide, Opoferzin from the Roots of *Ferula oopoda*. Khim. Prir. Soedin., **3**, 318 (1991); Chem. Abstr., **116**, 211114s (1982).

197. MURRAY, R.D.H.: Naturally Occurring Plant Coumarins. Fortschr. Chem. Organ. Naturstoffe, **35**, 199 (1978).

198. MURRAY, R.D.H.: Naturally Occurring Plant Coumarins. Fortschr. Chem. Organ. Naturstoffe, **58**, 83 (1991).

(*Received May 15, 1993*)

The Chemistry of Melanins and Melanogenesis

G. Prota, Department of Organic and Biological Chemistry,
University of Naples, Naples, Italy

Contents

I. Introduction

Some twenty years have passed since SWAN reviewed the chemistry of melanins for Vol. 31 of this series (*1*). Although some of the questions which he aired at that time have remained unanswered, our knowledge and appreciation of the field have increased dramatically in the intervening years (*2, 3*). Considerable steps forward have been made in practically all areas of melanin research, including the molecular biology of pigment-related genes (*4, 5*). Concurrent with these advances, research has begun to explore in earnest some fascinating phenomena at the interface between chemistry and biology, such as the origin of skin colour differences in man, the role of melanins in photoprotection and in a variety of genetic and acquired pigmentary disorders that are of great social concern. Some of these alterations are well known and include albinism, vitiligo, mongolian spots and melanoma. The latter is a tumor of the melanin producing cells whose increasing incidence, aggressive behaviour, and resistance to conventional therapeutic regimens represent a challenge to the whole scientific community.

It may therefore seem surprising that, in spite of the breadth and magnitude of the field, very few organic chemists have been involved in recent years in melanin research. This is probably due to the discouraging properties of melanins, especially the high degree of insolubility in all solvents which makes their isolation an almost impossible task. They also are highly inhomogeneous from the molecular viewpoint and lack well-defined spectral and physical characteristics; in consequence, the modern structural approach by spectroscopic techniques, particularly NMR and X-ray crystallographic analysis, which have been so successful in the elucidation of complex biopolymers, *e.g.* nucleic acids or proteins, are not applicable to the study of melanins. Indeed, it can be anticipated that most of what we know about the structure and origin of melanins has resulted from a biomimetic approach involving extensive studies of the chemical reactivity of the putative melanin precursors under biologically relevant conditions. Thus, bit by bit it has been possible to reconstruct in vitro the reaction pathway and to identify a number of biosynthetic intermediates which are pivotal for the understanding of melanin structure and properties. These and other approaches, particularly the identification of putative pigment precursors in vivo (*6*), have radically changed and expanded the chemistry of melanogenesis far beyond the time honoured Raper-Mason scheme.

It would not be realistic here to attempt to provide a detailed account of all these studies, nor would it be necessary since a comprehensive monograph covering all aspects of melanin pigmentation, including the biomedical ones, has appeared recently (3). The aim of the present contribution is therefore illustrative, with major emphasis on the structure and properties of epidermal melanins and related metabolites. It also aims at providing an overview of how far a purely chemical and mechanistic approach can go in unraveling the biochemical events underlying pigment formation in biological systems.

II. The Tangled Yarn of Melanin Research

Modern reviews in the chemistry of natural products are usually straightforward; the molecules, however complex, are often well defined both structurally and biogenetically. In the case of melanins, however, any discussion faces unavoidable difficulties due to the unusually long and entangled history of research that need to be considered before progress can be appreciated. Even the chemical meaning of the term melanin is controversial.

A. The Early Studies

Ethnic pigmentation was the natural phenomenon that provided the initial stimulus for melanin research (7). The early eighteenth century notion was that the colour of black skin was derived from bile, and ingenious theories were devised to explain how bile turned itself into the colouring matter of the skin. Later, however, it became apparent that the black colour was due to a granular insoluble pigment which was part of, and adherent to, the reticular membrane. This was confirmed by LE CAT (8) who extracted the skin of a black Ethiopian cadaver, and compared the isolated pigment with that of the choroid of the eye. Most remarkably for that time, he found that the two pigments (which he named ethiops) were not only closely similar to each other, but had also the same general features of the pigment from the ink of the squid, a deduction which proved to be correct.

After LE CAT's work comparative studies on animals were increasingly being performed by biologists. Similar pigments were then found in hair, melanomas, and other mammalian tissues including, notably, the substantia nigra of the brain. However, despite the promising start, research on the nature and origin of the pigmentary colour waned for the lack of experimental methods, and interest was therefore shifted to the biology and physiology of skin pigmentation. Perhaps a most important outcome

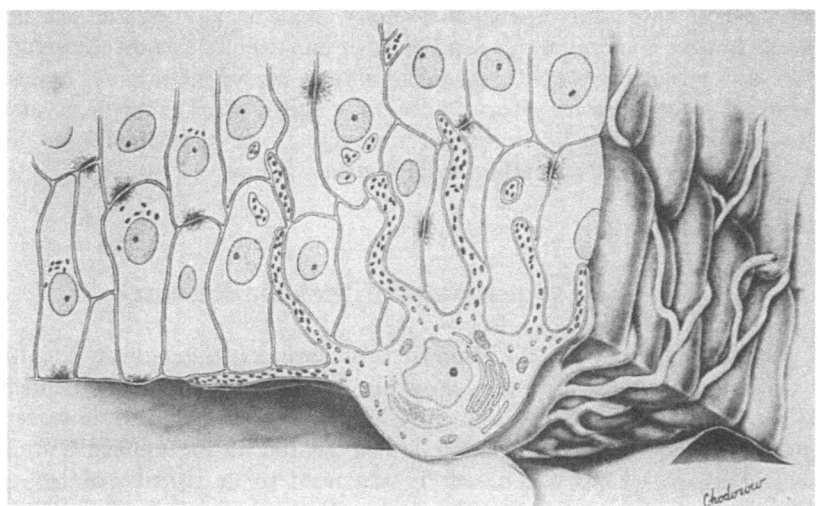

Fig. 1. Diagram showing the epidermal melanocyte with a group of keratinocytes with which it maintains functional contacts. By courtesy of Professor T.B. FITZPATRICK

of these studies was the demonstration that melanin formation takes place in highly specialized cells (Fig. 1), later described more precisely and named melanocytes (9).

B. Melanogenesis in Vitro

The first breakthrough in melanin research came at the turn of the century following the serendipitous discovery of tyrosinase in certain fungi, *e.g. Russula nigricans*, which blacken when injured and exposed to air. The melanogenic substrate was recognized by BERTRAND (10) as the phenolic amino acid tyrosine. When this was allowed to oxidize in the presence of a tyrosinase extract a black precipitate was eventually formed. A similar enzyme was subsequently found in extracts from many other plants and from the tissues of various invertebrates and vertebrates, including mammals (11). The fact that the general properties and elemental composition of the black precipitate resulting from the action of tyrosinase on tyrosine were in apparent agreement with those reported for melanins obtained from mammalian hair, eye, and melanomas, as well as sepia ink, implied that these natural pigments also were products of the tyrosine-tyrosinase reaction.

RAPER's fundamental studies on the early stages of the melanogenic reaction (*12*) are outlined in Fig. 2. He found that tyrosinase was an essential requirement for the initial conversion of tyrosine to dopa and then to dopaquinone, whereas the subsequent steps proceeded spontaneously without any specific enzymatic assistance. By stopping the oxidation of tyrosine at the dopachrome stage and allowing the red solution to decolorize under anaerobic conditions, RAPER was able to isolate (as dimethyl ethers) 5,6-dihydroxyindole (DHI) and a related compound in smaller amounts, identified as 5,6-dihydroxyindole-2-carboxylic acid (DHICA). At that time, the free phenolic indoles had neither been isolated nor synthesized. Accordingly, RAPER could do no more than make suggestions about the chemistry of the later stages of melanogenesis. Thus, in 1938 he wrote "Beyond the statement that, in the formation of this pigment (melanin) from dihydroxyindoles, quinone formation is a probable first step, it is not possible to go" (*13*).

In 1948 BEER, CLARKE, KHORANA and ROBERTSON (*14*) reported the synthesis of DHI and DHICA. They found that an exposure to air, especially in alkaline solution, DHI rapidly undergoes oxidative polymerization to give a dark insoluble melanin, whereas under similar

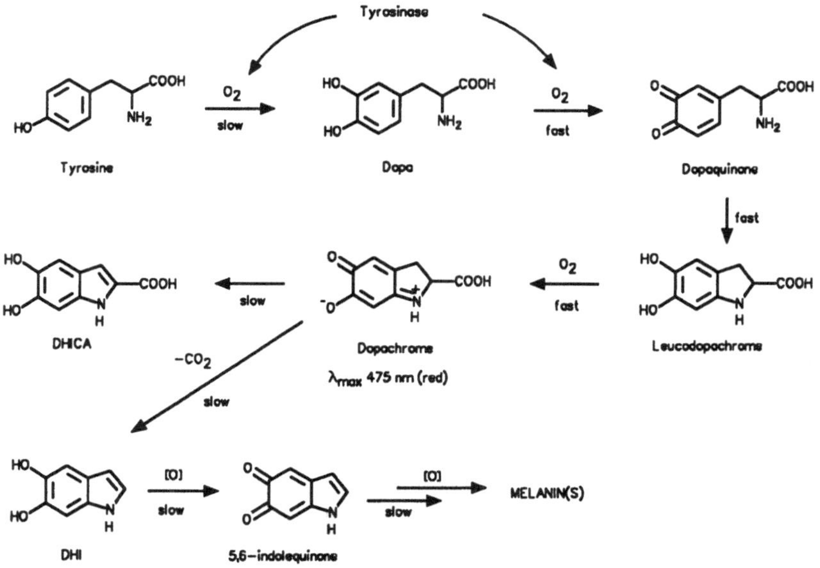

Fig. 2. RAPER's scheme of melanogenesis

conditions autoxidation of DHICA is much slower and yields a brown solution with no precipitation of melanin (*15*). From these experiments it was inferred that of the two putative indolic precursors suggested by RAPER only DHI was involved in melanogenesis. Such a view was confirmed by MASON (*16*), who found that oxygen consumption and carbon dioxide evolution during the enzymatic oxidation of dopa were consistent with DHI being the major ultimate precursor of the melanin polymer. Accordingly, he suggested that melanin was a homopolymer arising by repeated self-condensation of DHI. The proposed structure was later challenged by the analytical and degradative studies of NICOLAUS and PIATTELLI and coworkers (*17*), suggesting that melanin was rather a "poikilopolymer" derived by random coupling of various intermediary metabolites in the RAPER pathway.

MASON's and NICOLAUS's opposed theories of melanin structure were heatedly debated for several years. Eventually, the latter received the larger consensus from parallel in vitro studies with labelled precursors carried out at Newcastle and in other centres (for review see refs. *1, 17, 18*). At that time, however, the difference between natural and synthetic melanins was not fully appreciated and it was generally agreed that tyrosinase was the main, if not the sole enzyme involved in melanogenesis. As we shall see, the above turned out to be an oversimplification of the chemistry of melanin pigmentation.

C. Melanogenesis in Vivo

A most distinguishing feature of melanogenesis *in vivo* is the ability of epidermal melanocytes to produce a wide range of pigmentary colours, as seen for example in mammalian hair, which may be black, brown, yellow, reddish brown, or carroty red (*19*). At one time, these varieties of melanin were classified in two major groups, on the basis of their color and solubility properties: the black-to-brown, insoluble eumelanins (ευ = good) and the alkali-soluble pheomelanins (φαιος = dusky), ranging from yellow to reddish brown. Such a classification was supported by the genetic pattern of hair colour in mice indicating the existence of two separate but possibly interrelated pathways for eumelanin and phaeomelanin. Both types of pigments may occur together giving rise to spots, stripes and mottled appearance of many mammals and birds. Moreover, in the agouti mouse, the same melanocyte can apparently synthesize one type of melanosome after the other during a cycle of hair generation (*20*).

Although from these and other observations the existence of a switching mechanism was clear, for a long time it was thought that pheomelanins were chemically and biogenetically related to eumelanins; hence, the tendency to describe them collectively under the name melanin.

In 1958 FITZPATRICK and coworkers (21) re-opened the question, suggesting that pheomelanin formation could probably result from a deviation of the eumelanin pathway, due to the intervention of another substrate. They speculated that this could be tryptophan or the related metabolite, 3-hydroxykynurenine, which BUTENANDT had shown to be involved in the biosynthesis of the ommochromes. However, efforts to prove the involvement of tryptophan metabolites in the biosynthesis of pheomelanin in mice did not proceed very far and the proposed pathway was eventually abandoned.

An unexpected breakthrough came from investigations on a distinctive type of epidermal pigment which was first recognized in 1878 by SORBY (22) in acid extracts of carroty-red human hair. This pigment was eventually isolated in 1945 by FLESCH and ROTHMAN (23), who suggested a tentative formula $(C_{15}H_{20}N_2O_9)_2Fe$ and proposed the name trichosiderin ($\vartheta\rho\iota\xi$ = hair, $\sigma\iota\delta\epsilon\rho\sigma\varsigma$ = iron). In a subsequent study, however BARNICOT (24) obtained evidence that trichosiderin was probably an artifact since extraction of red hair with dilute alkali yielded an orange-yellow pigment that was converted into the red trichosiderin when heated with acid. Apparently unaware of BARNICOT's work, BOLDT (25) re-examined trichosiderin obtained by acid extraction and found that the crude pigment described by FLESCH and ROTHMAN was in fact a mixture of several components, some of which contained sulphur in addition to nitrogen (26).

In parallel studies at Naples (27), the more easily available red feathers of domestic fowl were used, which gave a similar mixture of red pigments. These, however, proved to be artifacts arising from unstable yellow precursors, as suggested by BARNICOT. Elemental analysis of the native pigments, isolated by mild alkaline extraction, confirmed the presence of sulphur, but no iron was found. The misleading name trichosiderins was therefore replaced by that of trichochromes (28). These proved to be biogenetically related to the polymeric pheomelanins, also found in red feathers. Both types of pigments arise from a deviation of the eumelanin pathway through the intervention of cysteine and/or glutathione (GSH), as evidenced by model experiments (27). Thus, despite the apparent differences in molecular size and general properties, eumelanins, pheomelanins, and trichochromes can all be embraced by one biogenetic scheme in which dopaquinone is the key intermediate (17, 29). We will return to this point later after reviewing the chemistry of the pigments.

III. Note on Terminology

The term melanin (μελας, black) is often broadly applied to cover almost any black or dark-looking pigment which accumulates in the form of insoluble granules in plant and animal tissues. To one who has been endeavouring to survey the literature, this wide usage has caused an immense amount of unnecessary reading and frustration. Therefore, we use and recommend here a more restrictive usage of the term to include only those pigments, such as eumelanins and pheomelanins, which are formed intracellularly by oxidation of tyrosine and related metabolites. From the chemical viewpoint, this is still not a satisfactory definition, but it may help to prevent confusion about exactly what constitutes a melanin. Thus, for example, the dark pigments found in higher plants, fungi, and bacteria are not melanins since they are usually extracellular in nature and arise by oxidation of nitrogen-free phenols, e.g. catechol and 1,8-dihydroxynaphthalene (17, 30). Accordingly, they are better referred to as allomelanins.

Another case in point concerns the melanic pigments which are associated with the stiffening and tanning of most insect cuticles (31). During this process, structural proteins are covalently modified and cross-linked by o-quinone and quinone methides arising by enzymatic oxidation of N-acetyldopamines (1) as well as N-β-alanyldopamine (2). The exact mechanism of polymerization of these substrates is still in question, but it seems clear that the resulting pigments are chemically distinct from typical eumelanins, though they also arise biogenetically from tyrosine.

(1 a) $R_1 = H$, $R_2 = CH_3$
(1 b) $R_1 = OH$, $R_2 = CH_3$ (2)

Similar problems occur with the structure of the non-dark melanins and consequently some disagreement exists as to what exactly constitutes a pheomelanin. As we have seen, an obvious difference between these pigments and eumelanins is the presence of sulphur, which derives biogenetically from cysteine. However, both in man and other mammals pheomelanin-looking pigments also occur which are devoid of sulphur and arise probably by peroxidative cleavage of 5,6-dihydroxyindole units of the eumelanin polymer (3). At present, there are no satisfactory tests to distinguish these bleached eumelanins or "oxymelanins" from authentic pheomelanins. Pigment colour and solubility in alkali are not specific; neither is the ultrastructure of the melanosomes which is often used for identification purposes (20).

A further complication in classifying melanins stems from the occurrence in epidermal tissues of certain pigments whose chemical and physical properties stand in between those of typical eumelanin and pheomelanin polymers. Analytical, degradative, and model studies, as yet imperfect, would suggest that such pigments arise biogenetically by copolymerization processes involving both eumelanin and pheomelanin precursors. However, whether they are heteropolymers or intimate mixtures of the two types of polymers is not yet clear (2, 3).

IV. Eumelanins

Although eumelanins are widespread in the animal kingdom (17), only a few of them have been isolated and chemically characterized. One of these is sepiomelanin, easily accessible in large quantity from the ink sac of the cuttlefish *Sepia officinalis*. A commercial preparation of this pigment is now available which finds increasing use in sunscreen formulations (3). Other eumelanins which have been studied in some detail include those from bovine eyes, melanoma bearing B-16 mice and black hair from human subjects, horses, poodles, and ox tails.

A. Protein Content

In their native form, eumelanins are generally associated with protein, as evidenced by amino acid analysis. The protein can be removed, with difficulty, but the untoward properties of the pigments make it impossible to establish if and how they are linked together. Conjugation could occur either by addition of cysteine residues to the pigment polymer to form a sulphur-linked melanoprotein, or by oxidation of a melanosomal protein, containing a terminal tyrosine residue. The former mechanism would be in keeping with the presence of some sulphur in protein-free eumelanins and the formation of cysteic acid and taurine by peracetic oxidation of sepiomelanin which had been treated extensively with 6M HCl (32). However, the same data would also be in keeping with the incorporation

of non-protein thiols, especially GSH, to the growing pigment polymer
(*33, 34*).

(3a) R=H
(3b) R=COOH

(4a) R=H
(4b) R=COOH

(5)

(6)

(7)

The position of the sulphur bond is not known, although nucleophilic
addition of –SH to C-4 of 5,6-indolequinone units appears to be the most
likely. This is supported by model studies (*35, 36*) showing that GSH
efficiently reacts with enzymatically generated 5,6-indolequinone (**3a**) and
its 2-carboxyl analogue (**3b**), to give the C-4 linked adducts (**4a**) and (**4b**),
respectively. At physiological pH, *e.g.* 6.8, GSH also reacts with dopachrome, but the resulting adduct (**5**) is rapidly oxidized to (**6**) which then
rearranges with loss of carbon dioxide to give (**4a**). If cysteine is used in
place of GSH, a different reaction takes place leading to a yellow non-
aminoacidic product with an absorption maximum at 422 nm. The
structure of this compound is uncertain, but the corresponding adduct
arising from reaction of dopachrome with cysteine ethyl ester proved
to be (**7**), containing the new 1,2-dihydro-3H,8H-pyrrole[2,3-h]benzo-
thiazine ring system.

There is no direct evidence for the alternative mode of conjugation of
the protein through a peptide linkage, but it is noteworthy that tyrosine
containing peptides can be oxidized by tyrosinase. This was first shown
by BU'LOCK and HARLEY-MASON (*37*) using, as a model peptide, tyrosyl-
glycine. On oxidation with mushroom tyrosinase, it gave a red solution
with a dopachrome-like chromophore that eventually turned brown
without deposition of pigment. When kept under nitrogen, the red
solution slowly decolourized, forming apparently a dihydroxyindole
peptide (Fig. 3). The latter could presumably be oxidized to the quinone
which might then be able to copolymerize with 5,6-indolequinone derived
from unbound tyrosine.

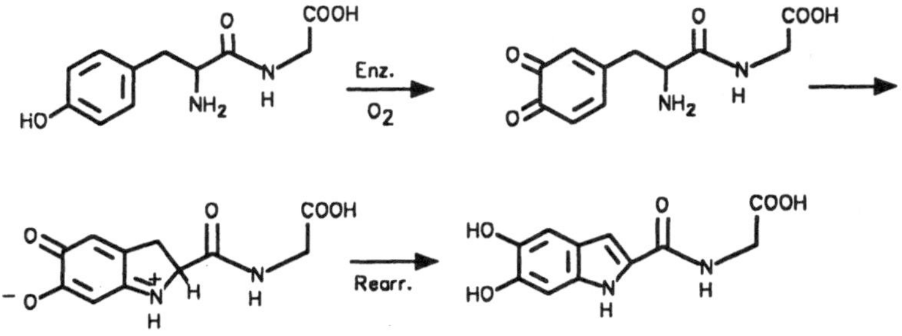

Fig. 3. Tyrosinase-catalyzed oxidation of tyrosinylglycine

Similar results have been obtained by YASUNOBU et al. (38) and more recently by ROSEI et al. (39) on a number of tyrosine containing peptides, including Leu-enkephalin and Met-enkephalin. Depending on whether tyrosyl residues are at the N-terminus, at the C-terminus or in the middle, a dopachrome absorption curve (in the former case) or a dopaquinone absorption pattern (in the latter cases) were observed, consistent with the proposed oxidation scheme.

Notwithstanding all these studies, the question of whether natural eumelanins are melanoproteins in the usual meaning of the term, *i.e.* covalently bound to protein, remains elusive. In a relevant study at Lausanne (40), it was found that native sepiomelanin loses in hot 6M HCl most of the "proteic" component during the first 15 mins, the remaining part being removed in about an hour. Analysis of the solubilized fraction revealed a mixture of aminoacids (11%) along with little amounts of glucosamine. It seems therefore that a considerable part of protein is loosely bound to the pigment granules which raises some doubts about the commonly held view of sepiomelanin being a melanoprotein.

B. Isolation and Analysis

Procedures for extraction of eumelanins from sources such as hair, feathers, eyes, melanomas and cephalopod ink usually involve destruction or removal of all other components of the tissue by prolonged digestion with conc. HCl at room temperature or by boiling with 6M HCl (17, 41). Of these two options, the latter should be avoided if possible since it may cause some changes in the pigment structure and composition. This has been carefully documented by BENATHAN and WYLER (40), who found that on heating in 6M HCl sepiomelanin as well as other eumelanins undergo extensive decarboxylation accounting for as much as 23% weight loss of the original material.

Following these studies, increasing attention is currently being devoted to develop mild procedures for obtaining native eumelanins. In some favourable cases, *e.g.* sepiomelanin, eye-melanin, these can be isolated with minimal alteration by mechanical separation of the pigment granules followed by a short treatment with 0.5 N HCl at room temperature and extensive sonication in deionized water (*40*). Likewise, melanosomes from melanoma tissues have been purified by sucrose density gradient ultracentrifugation after digestion with detergents or proteolytic enzymes, *e.g.* pronase and papain, at neutral pH (*3*). Whether these mild treatments efficiently remove proteins and other extraneous materials incorporated or bound to the pigment granules is not clear, but this is undoubtedly a minor drawback with respect to the alternative

Table 1. *Elemental Analysis of Some Natural Melanins*

Source	% C	% H	% N	S%	C/N	Ref.
Sepia ink[a,b]	50.8	3.1	8.4	0.04	7.0	(*51*)
Sepia ink[c]	64.7	2.5	9.0	0.3	8.4	(*17*)
Sepia ink[d]	60.9	3.4	8.4	0.3	8.4	(*61*)
Sepia ink	56.0	3.2	8.9		7.3	(*40*)
Sepia ink[c]	57.7	2.6	6.2		10.9	(*40*)
Sepia ink[c]	52.2	3.4	7.9	0.2	7.7	(*42*)
Squid ink[d]	60.9	3.1	6.0		11.8	(*58*)
Octopus ink[d]	59.0	3.3	8.0		8.6	(*58*)
Bovine eye [c]	53.6	2.7	7.5		8.3	(*40*)
Bovine eye[d]	54.9	4.2	9.3	0.8	6.9	(*86*)
Bovine eye[d]	60.4	4.6	8.7	0.9	8.1	(*58*)
B-16 Melanoma[d]	64.5	5.5	7.1		10.6	(*58*)
B-16 Melanoma[c]	52.3	4.5	7.3	0.6	8.4	(*42*)
Mouse hair[c]	56.5	4.6	7.6	0.8	8.7	(*42*)
Horse hair[c]	64.6	5.1	6.0	1.3	12.6	(*58*)
Dog hair[c]	61.2	4.0	5.1		14.0	(*58*)
Black human hair[c]	61.1	4.6	8.4	3.7	8.5	(*58*)
Black human hair[e,f]	66.6	8.0	5.2	1.1	14.9	(*51*)
Bovine hair[c]	52.8	4.0	8.3	0.9	7.4	(*86*)
Ox hair[c]	65.5	5.5	5.4	1.6	14.1	(*58*)

[a] Equilibrated with 3.4 M KCl
[b] Repeated analyses gave: C ± 1.0%, H ± 0.3%, N ± 0.2%, S ± 0.1%
[c] Treated with boiling HCl
[d] Treated with cold conc. HCl
[e] Acid-treated and equilibrated with 3.4 M KCl
[f] Repeated analyses gave: C ± 0.5%, H ± 0.1%, N ± 0.15%, S ± 0.03%

of producing a structurally modified product as in the drastic acid procedure.

Given the problems encountered with the isolation of the pigments, elemental analyses have thus far only provided limited information about the actual composition of the pigment polymer (Table 1). Pronounced variations have been reported in the analytical values of eumelanins from different sources, but whether they arise from intrinsic differences in the structure of the pigment polymer or from artifactual alterations during the work-up procedure is not clear. This latter view seems more likely, since repeated analyses of the same pigment, e.g. sepiomelanin, are often significantly divergent. Other factors accounting for the lack of reproducibility include the presence of protein and other impurities, especially metal ions, and the hydration state of the sample. In view of these and other limitations, use of the molar ratio of carbon to nitrogen (C/N) of the pigment is preferred for characterization purposes (42). As a rule, C/N ratios lower than 8.1, corresponding to that of a 5,6-dihydroxyindole, are indicative of acid degradation of the pigment backbone with loss of carbon dioxide, while higher values suggest the presence of associated protein and/or of C-9 units, such as DHICA.

C. Synthetic Melanins

As mentioned earlier, considerable work has been done on synthetic melanins obtained by oxidation of known precursors under conditions which are modelled upon proven or speculative biosynthetic pathways. They are named after the compound from which they are prepared, e.g. dopamelanin, dopamine melanin, 5,6-dihydroxindole melanin, cysteine-dopa melanins, etc. The exact conditions under which the oxidation is performed are extremely important: Variations, even minor ones, can have a marked effect on the reaction sequence and, hence, on the structure and properties of the final product. Unfortunately, methods for preparation of synthetic melanins vary according to the habit of the authors and a standardized procedure, though highly desirable, has never been developed. In the case of dopamelanin, for example, the conditions used include (i) enzymic oxidation with tyrosinase in air for 6 hr up to 3 days at room temperature, or (ii) with pure oxygen for 4–6 hr at 25 or 37 C, and (iii) autoxidation in phosphate buffer or ammonia solution at pH 8 with or without catalase. Although these melanin preparations look superficially alike, they are by no means similar, as evidenced by an overwhelming body of analytical, degradative, ESCA and NMR data (40–45).

One major difference is in the content of carboxylic groups which is significantly higher in autoxidative dopamelanin than enzymically produced dopamelanin. This conflicts with the statement of SWAN and WAGGOTT (46) that enzymatic dopamelanin prepared in vitro in the way they described is not much different structurally from autoxidative melanin. It may be noted however that SWAN's enzymatic and autoxidative dopamelanins are rather unique in that they were both obtained by prolonged oxidation, i.e. for 3 days, followed by treatment with acids. It is possible that extensive degradation of the pigment polymer may occur as a result of such drastic treatments, which eventually make the two pigment preparations chemically similar.

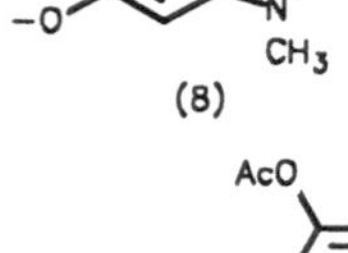

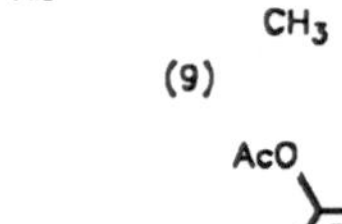

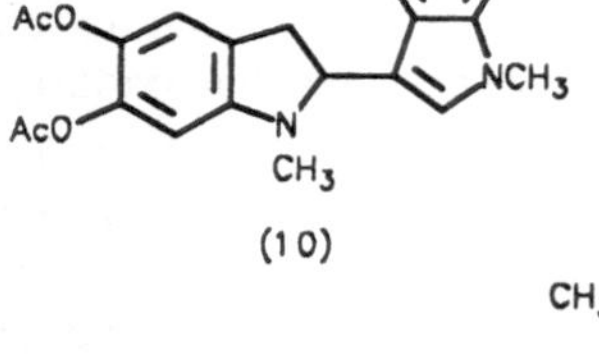

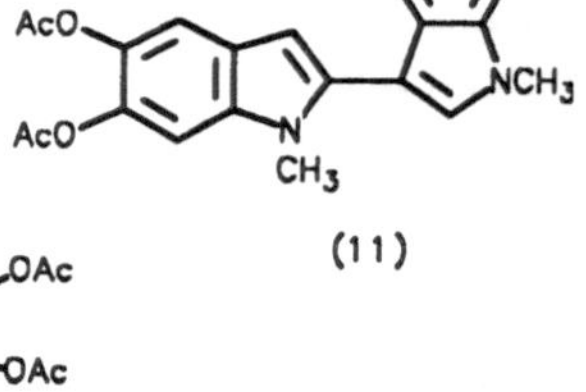

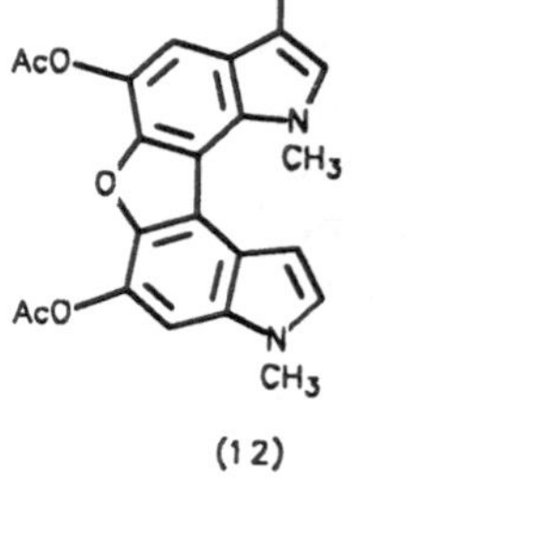

At one time it was thought that the black insoluble pigment derived by acid catalyzed polymerization of adrenochrome (8) via 1-methyl-5,6-dihydroxyindole (9) could provide a useful model in studies of melanogenesis, and in particular of the mechanism of polymerization of 5,6-dihydroxyindoles (47). However, a reexamination of the structure of "adrenalin black" (48) has shown that the pigment consists mainly of dimers and trimers such as (10), (11) and (12) arising evidently by enamine-imine condensation of (**9**) rather than by oxidative coupling at positions 3 and 7, as previously suggested (*47*).

D. Free Radicals and Redox Properties

Historically, eumelanins were the first natural products which were studied by EPR spectroscopy and found to be paramagnetic (*49*). Considerable efforts have been made to correlate the free radical properties with some of the alleged functions of melanins in the physiology of vision and in photoprotection, but the results are controversial (*50*).

Reported EPR spectra of eumelanins, be they natural or synthetic, are remarkably similar and include a single rather featureless signal with a line width of about 4–6 G and a g-value close to 2.004. No hyperfine coupling is detected and the spin concentration is very small, usually within the range of $4–10 \times 10^{17}$ spins/g. Semiconductor models and charge transfer complexes through the stacked monomer units of the eumelanin polymer have been proposed to account for such an unusual type of stable organic radicals (*52–54*), but most workers favour BLOIS's interpretation that the unpaired electron is due to a small fraction of semiquinonoid units which are deeply embedded into the polymer (*18*). This is supported by the fact that the concentration of the intrinsic melanin radicals can be increased reversibly under a variety of experimental conditions, *e.g.* alkaline pH, metal complexation, which are likely to affect the redox equilibrium:

$$Q + QH_2 \rightleftharpoons 2Q^{-\cdot} + 2H^+,$$

where Q and QH_2 represent 5,6-indolequinone and 5,6-dihydroxyindole units of the pigment polymer, respectively. It is clear, however, that this interpretation depends on whether one relies on the quinone polymer model to the structure of eumelanins, which is not generally accepted (*2*).

More recently, emphasis has been shifted to the ability of eumelanins to generate an additional pool of free radicals following irradiation with UV and visible light under nearly physiological conditions (*55*). These transient radicals consist of two pools, one characterized by a low yield (ca. 1–2%) and a decay time of a few seconds, and the other with a decay time of a few milliseconds, that accounts for about 50% of the signal intensity. Under continuous irradiation, the yield of this latter component is 50–100 times that of the former and the photoexcitation process shows the characteristics of an intersystem crossing mechanism.

Work on aerated samples has revealed an additional complexity deriving from concomitant photooxidation of the pigment polymer and formation of superoxide anion and hydrogen peroxide (56). The action spectrum of oxygen consumption is closely related to that of free-radical production, increasing with pH of the medium. The phenomenon is further enhanced by the addition of NADH or other reducing agents which evidently convert the pigment to a more oxidizable form. This was originally ascribed to a direct two-electron reaction but new evidence points to a more plausible one-electron reduction of oxygen to super-oxide, followed by dismutation and/or further reduction by melanin, yielding hydrogen peroxide (55).

E. Carboxylic and Phenolic Groups

From the early studies of NICOLAUS and PIATTELLI (57), two types of acidic functions are known to be present in eumelanins, phenolic and carboxylic. The latter were determined by titration, decarboxylation, or by esterification with methanolic hydrogen chloride followed by analysis of the methoxyl content. In the case of sepiomelanin, the difference between the methoxyl value of specimens treated with methanolic hydrogen chloride (6.0%) and with diazomethane (18.8%) gave a content of phenolic hydroxyl groups of 13.1%. Similar results were obtained for other pigments investigated, including enzymatic dopa and DHI melanin. This has been taken to mean that the pigments are not fully quinonoid but are approximately halfway between the quinonoid and diphenolic stages (46). However, a reliable estimation of the relative proportion of oxidized and reduced indole units within the polymer is difficult, since functionalities other than phenolic and carboxyl groups may react with diazomethane, as evidenced by the formation of methylenedioxy groups and N-methylated pyrrole rings (1, 17).

BENATHAN and WYLER (40) and, more recently CHEDEKEL and ZEISE (59) used titration to determine the ratio of moderately acidic (carboxylic) to weakly (phenolic?) acidic functions. The consistent findings indicate that the ratio of these acidic groups in native eumelanins, e.g. sepio-melanin, is 1.1. After removal of the protein with acid it drops to about 0.8, due probably to the loss of some carboxylic groups. Significant differences were also observed between enzymatic (0.5) and autoxidative (1.7) dopamelanin, the latter having the highest proportion of carboxylic groups.

This is in accord with isotopic and degradative studies by SWAN and WAGGOTT (46). These showed that during melanogenesis in vitro hydrogen peroxide is generated, which attacks the 5,6-indolequinone units of the pigment to give rise to carboxylated pyrrolic units (Fig. 4). The

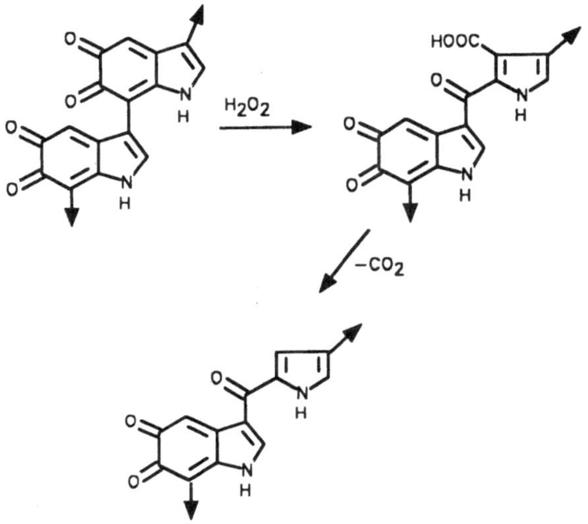

Fig. 4. Proposed model of peroxidative cleavage of 5,6-dihydroxyindole units in the eumelanin polymer. Adapted from ref. (1)

same mechanism also seems to be operative in vivo and would account for the occurrence of similar carboxylated pyrrolic units in natural eumelanins (32).

Recently, CRESCENZI et al. (60) examined the stability of dopamelanin to autoxidation at neutral pH. Unexpectedly, they found that the pigment is remarkably unstable, the carboxyl content increasing with oxygenation time from 4.8 (4 h) to 8.9% (24 h). This suggests that during the synthesis of dopamelanin some hydrogen peroxide can also be generated by redox exchange processes of oxygen with the pigment being formed. And indeed, in control experiments, it was found that the longer the oxygenation time the higher the carboxyl content of the nascent pigment, as evidenced by acid decarboxylation (Fig. 5). Moreover, when catalase was added to the incubation mixture, the carboxyl content of the dopamelanin was significantly reduced. Following these observations, a representative set of dopamelanins prepared according to various laboratories was also analyzed, the various dopamelanins were shown to differ from each other with respect to carboxyl content (see Fig. 5).

These results strongly suggest that the acidity and, hence, the structural integrity of eumelanins, be they natural or synthetic, may be

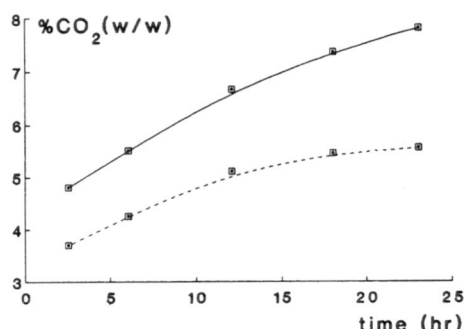

Fig. 5. Carboxyl content of enzymatically prepared dopamelanin as a function of the incubation time: dashed line, in the presence of catalase; solid line, in the absence of catalase

affected to a considerable extent by circumstantial peroxidative clevage of the labile 5,6-dihydroxyindole units. This makes it difficult to compare analytical data from different laboratories and to draw conclusions regarding the structure and composition of natural synthetic melanins. Hopefully, awareness of the instability of the pigment to oxygen as well as to acids will help develop more adequate procedures for obtaining natural and synthetic melanins with minimal structural alterations.

F. Chemical Degradation

The most complete study of chemical degradation of eumelanins was made in the 1950s and 1960s by NICOLAUS, PIATTELLI and their associates (*32, 61*). They found that oxidation of sepiomelanin with potassium permanganate or with peracetic acid leads, besides much oxalic and oxamic acids, to 2,3,5-pyrroletricarboxylic acid (PTCA, **13**) in about 2% yield. Other minor fragments, which could be identified, include the isomeric 2,3,4-pyrrole-tricarboxylic acid (**14**), pyrrole-2,3-dicarboxylic acid (PDCA, **15**), and 2,3,4,5-pyrroletetracarboxylic acid (**16**). Although the yields of these latter acids were exceedingly low, they appeared to have some structural meaning. Thus, when decarboxylated sepiomelanin was oxidized with permanganate, the yield of PDCA increased while that of PTCA decreased; moreover, pyrrole-2,4- (**17**) and pyrrole 2,5-dicarboxylic acid (**18**) were also obtained which are absent among the oxidation products of the non-decarboxylated pigment.

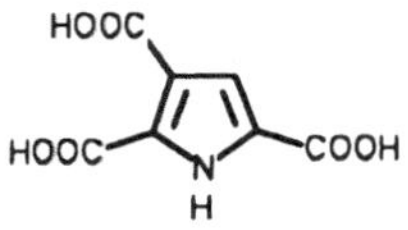

(13) 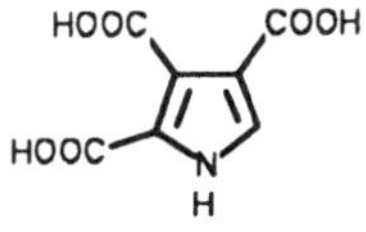   (14) 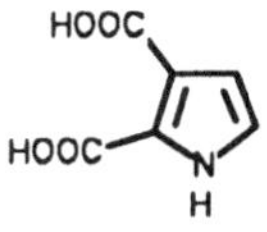   (15)

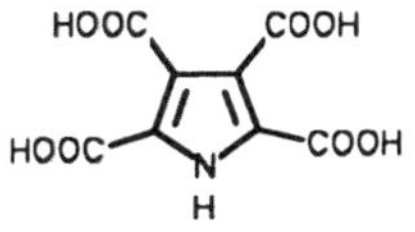

(16) 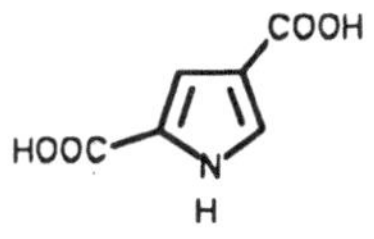   (17) 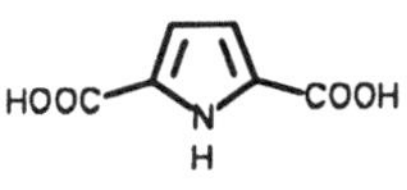   (18)

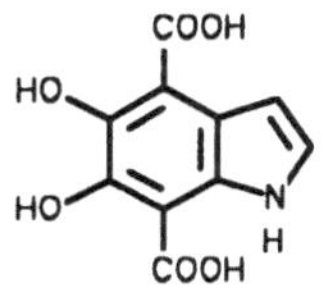

(19) 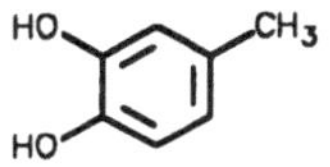   (20)

The origin of these pyrrolic acids was interpreted in terms of oxidative breakdown of various types of 5,6-dihydroxyindole units in the pigment backbone (Fig. 6). This was substantiated by alkali fusion of sepiomelanin, which afforded, albeit in trace amounts, DHI, DHICA, 5,6-dihydroxyindole-4,7-dicarboxylic acid (19), and 4-methyl catechol (20). The latter, which was not obtained by alkali fusion of DHI melanin, was ascribed to the presence of dopachrome units in the natural pigment (61). Moreover, when the pigment was boiled with 4% aqueous NaOH solution, PTCA was obtained in some 1% yield, arising presumably from hydrolysis of terminal, carboxylated pyrrole units linked through a carbonyl group to other structural units. Similar results were obtained by degradation of other natural eumelanins as well as some synthetic pigment preparations, including DHI melanins (58, 61–63).

As first noted by MASON (16), however, the melanins used in these studies were usually obtained under rather harsh conditions and the exceedingly low yields of the degradation products identified make it difficult to assess their importance as far as the bulk structure of the pigment polymer is concerned. More recently, ITO (42) found that the pattern and relative yields of pyrrolic acids obtained by permanganate oxidation of eumelanins at alkaline pH are significantly affected by the

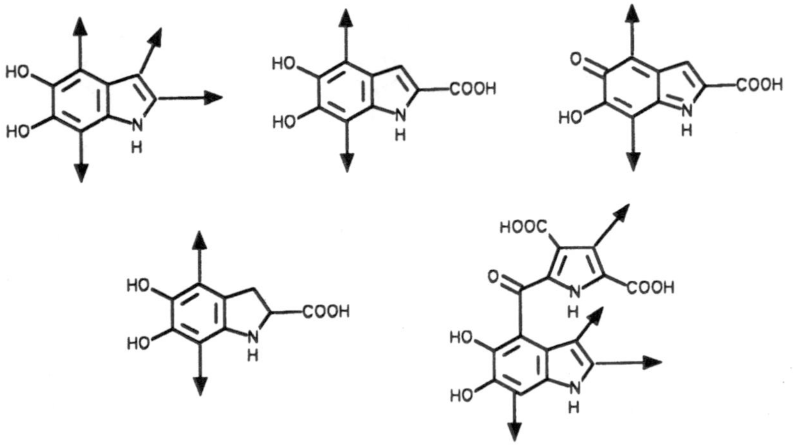

Fig. 6. Suggested structural units of sepiomelanin [ref. (*17*)]

artifactual formation of the tetracarboxylic acid. More reproducible results could be obtained by permanganate oxidation in 1M sulphuric acid which gave PTCA in similar yields (30 µg/mg) along with trace amount of PDCA only. In another relevant study at Naples (*64*) it was found that mild treatment of sepiomelanin or dopamelanin with sodium borohydride results in the formation of some DHICA, arising presumably from the incorporation of some pigment precursors in the melanin granule.

From time to time, a number of other approaches to the chemical degradation of eumelanins have been explored, but little progress has been made (*3*). BENATHAN and WYLER (*40*), for example, found that treatment of eumelanins with periodate results initially in the formation of a soluble polymeric product which on further oxidation gives rise to oxalic and oxamic acid along with trace amounts of pyrrolic acids. More extensive degradation was observed when periodate was used in conjunction with a catalytic amount of permanganate (Lemieux-von Rudloff procedure), but again the reaction products gave mainly oxalic and oxamic acids, along with trace amounts of some pyrrolic fragments. Three of these, formed in less than 1% yield, were identified by GC-MS as the methylated derivatives (**21**), (**22**) and (**23**) after methylation with diazomethane.

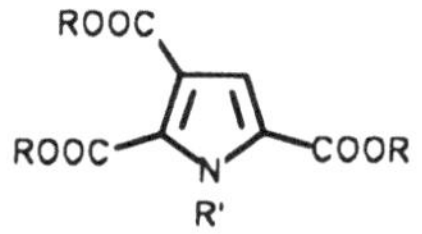

(21 a) R=R'=Me

(21 b) R=Me; R'=H

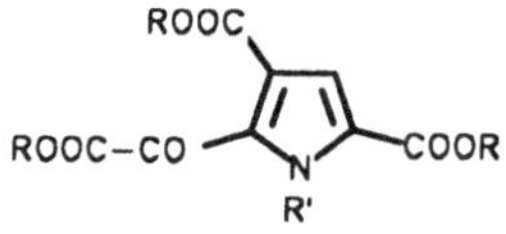

(22) R=R'=Me

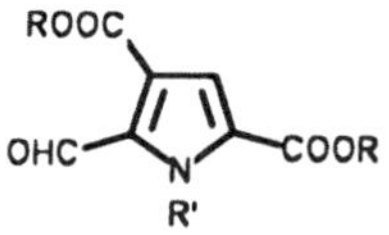

(23a) R=R'=Me

(23b) R=Me; R'=H

G. Biomimetic and Biosynthetic Studies

Since biosynthetic precursors are known, another way to inquire into the chemistry of melanins has been to study their reactivity under biologically relevant conditions. This approach, which was first exploited with remarkable success by RAPER, has recently yielded considerable insight into the mode of formation and structure of the eumelanin polymer that otherwise would have been difficult to obtain. An overview of the progress so far achieved will be given later. Here we elect to discuss two critical issues in the biosynthesis of melanins, *i.e.* the rearrangement of dopachrome and the oxidative polymerization of 5,6-dihydroxy-indoles.

1. Rearrangement of Dopachrome

a. Mechanism of Reaction

Until recently, two possible mechanisms were considered to account for the isoelectronic conversion of dopachrome to DHI (65) (Fig. 7). The first involves decarboxylation to give the indolenine (24) followed by its subsequent tautomerization to DHI (Pathway a). Alternatively, hydrogen shift from position 3 could give a quinone-methide (25) which would then undergo decarboxylation to DHI (Pathway b). That the latter was probably correct could be inferred from the isolation of the stable quinone methide (27) by rearrangement of α-methyldopachrome methylester (26), in which both decarboxylation and H-shift from the 2-position are prevented (66–68).

In a confirmatory study (69), the rate constants for the rearrangement of [2-^{2}H]- and [3, 3-^{2}H$_2$] dopachrome were measured and found to be 0.055 and 0.0073 min^{-1}, respectively. The primary isotope effect (k_H/k_D = 8.2) observed for [3, 3-^{2}H$_2$] dopachrome indicates that base catalyzed removal of H-3 is the first step of the reaction, consistent with the intermediacy of the quinone methide.

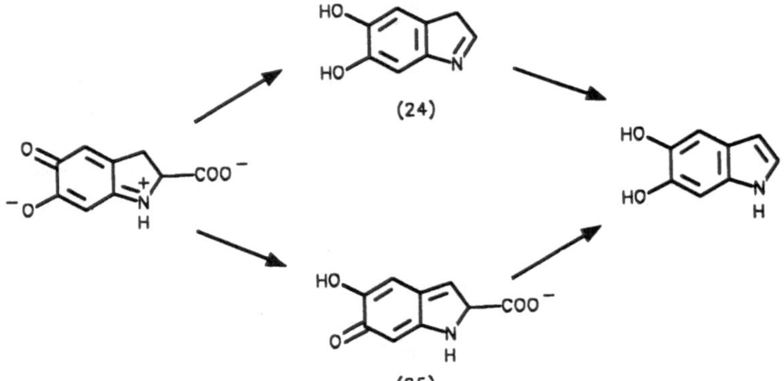

Fig. 7. Alternative mechanisms of rearrangement of dopachrome to DHI

Regarding the mode of rearrangement of dopachrome, analytical data from various laboratories (69–71) have shown that in the whole range of pHs from 3 to 8.5 the yield of DHI vs. DHICA is about 95 to 5. This requires that synthetic dopamelanins prepared by tyrosinase-catalyzed oxidation are made up mainly of DHI-derived units, as held by MASON (16). In consequence, the carboxyl content of dopamelanins must to a large extent be due to peroxidative cleavage of indole units, rather than to a random incorporation of early carboxylated biosynthetic intermediates (e.g. dopa, dopachrome and leucodopachrome), as contended by supporters of the poikilopolymer model of melanin structure (1).

An additional mechanism accounting for the higher carboxyl content of autoxidative dopamelanin has been offered by STRAVS-MOMBELLI and WYLER (72). These authors found that on exposure to oxygen or hydrogen peroxide at neutral or alkaline pH, dopachrome undergoes ring cleavage and epoxidation to give, inter alia, the lactonic acids (28) and (29), isolated as the dimethylesters (Fig. 8). However, whether such units are actually present in autoxidative melanin is still open to question.

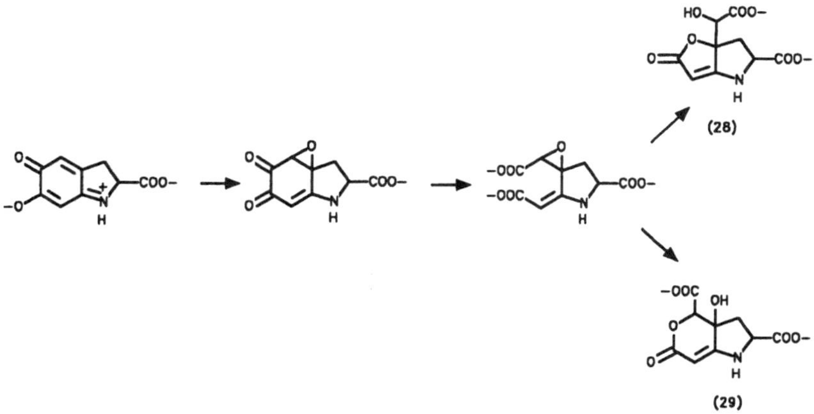

Fig. 8. Autoxidation of dopachrome. Adapted from ref. (72)

b. Regulatory Factors

Certain metal ions, especially copper, zinc and iron, have long been known to be involved in melanin pigmentation (3). Very high levels of these metals and others, *e.g.* manganese, have been found in a variety of melanized structures, including the choroid of the eyes (73), red or black hair (74), pigmented moles (75), and isolated melanosomes from hamster and human melanomas (76).

The possibility that metal ions could catalyze some steps in the biosynthesis of melanins was first suggested by HARLEY-MASON and BU'LOCK (77). They reported that in the presence of zinc acetate dopachrome rearranges very rapidly to give DHI. Following these studies, zinc has often been included in the Raper-Mason scheme as a natural promoting factor for the conversion of dopachrome to DHI (1). However, the concentration of zinc acetate used in these experiments was far from being biomimetic, and, as admitted by the authors, only in one experiment a 30% yield of DHI was obtained. This has been ascribed (78) to the ability of the metal ion to catalyze two sequential reactions, *i.e.* the rearrangement of dopachrome and the subsequent oxidation of the resulting indole(s). In fact, under rigorously anaerobic conditions, the zinc catalyzed reaction gives mainly DHICA rather than DHI (Fig. 9). A similar directional effect is observed with other metal ions, including copper and iron, which proved to be by far more effective than zinc in catalyzing the non-decarboxylative rearrangement of dopachrome (79). As a rule, the greater the accelerating effect of the metal ion, the higher the extent of conversion to DHICA.

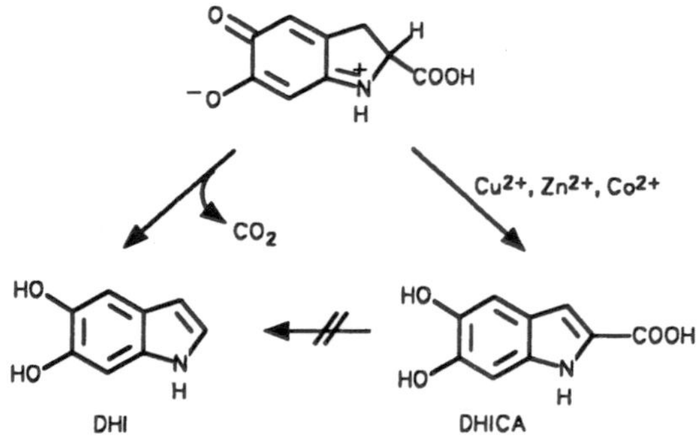

Fig. 9. Effect of metal ions on the rearrangement of dopachrome. Under physiological conditions, no conversion of DHICA to DHI takes place

Recent studies beginning with those of KORNER and PAWELEK (80) have revealed the presence in melanocytes of a melanosomal protein different from tyrosinase, which has the same ability as metal ions to catalyze the rearrangement of dopachrome to DHICA. The enzyme catalyzed reaction is highly stereospecific for L-dopachrome, is unaffected by metal chelators and has an optimal pH of about 6.8. Different names have been proposed for this enzyme, *i.e.* dopachrome conversion factor (80, 81), dopachrome oxido reductase (82, 83), dopachrome isomerase (84) and dopachrome tautomerase (85), the last being the one generally accepted.

It is of interest that another example of an enzyme acting on dopachrome has been described in insects (68). This has been named dopaquinoneimine conversion factor and has the remarkable ability to catalyze the decarboxylative rearrangement of dopachrome to DHI rather than DHICA. A similar enzyme has recently been found in the ink of the cuttlefish *Sepia officinalis* (87), but whether it is involved in the biosynthesis of sepiomelanin awaits further investigation.

The finding that the rearrangement of dopachrome is under regulatory control has prompted several studies aimed at evaluating the extent of incorporation of DHICA into the melanin polymer. Using a combination of analytical and degradative techniques ITO (42) presented evidence that DHICA-derived units comprise only 10% of enzymatic dopa melanin but as much as a half of intact eumelanins from sepia, black hair and B-16 melanomas. In a more recent study (88), the incorporation of DHICA into the melanin polymer was directly probed *in vivo* by inoculating intravenously L-3,4-dihydroxy[1-^{14}C]phenylalanine and L-3,4-dihydroxy[3-^{14}C]phenylalanine in melanoma-bearing mice. The tumour melanins were then carefully purified, analyzed for total radioac-

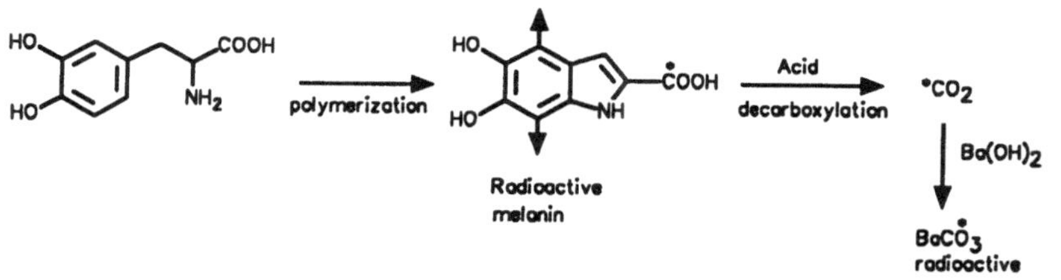

Fig. 10. 1-^{14}C-Dopa as a probe to assay the incorporation of DHICA into melanoma melanin

tivity and then subjected to chemoselective decarboxylation with boiling hydrochloric acid, which affects only those carboxyl groups linked to aromatic-type units, i.e. to DHICA-derived units (Fig. 10). The reference melanin derived from L-3,4-dihydroxy[3-^{14}C]phenylalanine yielded no radioactive carbon dioxide, whereas the pigment originating from L-3,4-dihydroxy[1-^{14}C]phenylalanine liberated on decarboxylation a substantial fraction of the total radioactivity, from which an incorporation of DHICA of more than 20% could be estimated.

2. Polymerization of Eumelanin Precursors

a. 5,6-Dihydroxyindole

Enzymic oxidation of DHI leads in the early stages to a transient purple pigment, known as melanochrome, with a broad absorption maximum at 540–560 nm. Since it was first described by MASON (89) in 1948, the origin of this chromophore generated a tremendous amount of work, in view of its relevance for the understanding of the structure and mechanism of formation of eumelanins. The early studies were largely speculative and were based on the assumption that the polymerizing species, 5,6-indolequinone, could possibly behave both as a quinone and as an indole, the anionoid centre in the pyrrole ring of one molecule linking up to a cationoid position of another. BU'LOCK and HARLEY-MASON (90) showed that such reactions were feasible by model experiments with simple quinones and indoles (Fig. 11). The dominant products were the intensely blue indolylquinones (30), indicating the involvement of the 3-position of the indole ring. If such position was blocked, the reaction proceeded with difficulty to give colourless products, if any.

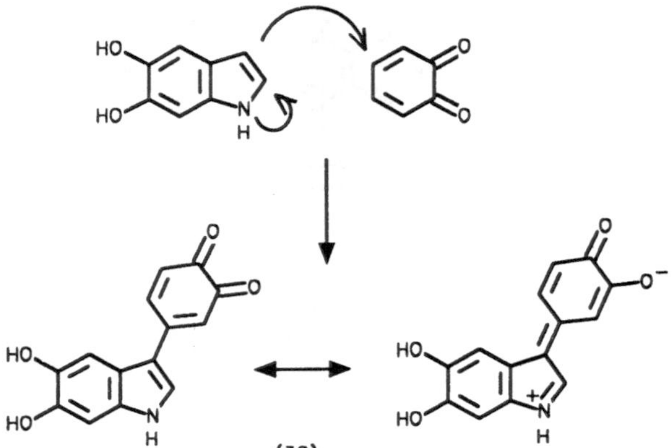

Fig. 11. Michael addition of indole to 1,2-benzoquinone

In further studies at Cambridge (*91, 92*) and other centres (*93*) it was found that oxidation of 1-, 2-, 4- and 7-methyl-5,6-dihydroxyindoles affords melanic polymers insoluble in the common organic solvents but slightly soluble in pyridine, as in the case of DHI. On the other hand, 3-methyl- and 4,7-dimethyl-5,6-dihydroxyindole give highly coloured products soluble in ethanol. From consideration of all the above experiments, it was concluded that the 3-position was essential for formation of a true melanin, together with either a 4- or a 7-position free, though all positions (2, 3, 4 and 7) were probably involved to some extent in the building up of a three-dimensional polymer.

The suggested mode of polymerization was in some disagreement with the tracer studies by SWAN (*1*) and by KIRBY and OGUNKOYA (*94*) which indicated that in the oxidation of dopa to melanin each of the various positions of the side chain and benzene ring were involved to a similar extent. However, no firm conclusion about the mode of polymerization of DHI could be drawn from these studies. In a reexamination of the problem at Naples (*95*), a procedure was eventually developed for obtaining melanochrome in an amount sufficient for chemical investigation. This involved autoxidation of DHI at neutral pH in the presence of zinc ions acting as catalyst. Reduction and subsequent acetylation of the blue melanochrome thus formed afforded a mixture of oligomers of DHI, the major of which proved to be the symmetrical dimer (**31**) arising by oxidative coupling of DHI at the 2-position.

In further studies (*96*), the oxidation of DHI was performed under the usual conditions of melanogenesis *in vitro, i.e.* with the enzyme tyrosinase

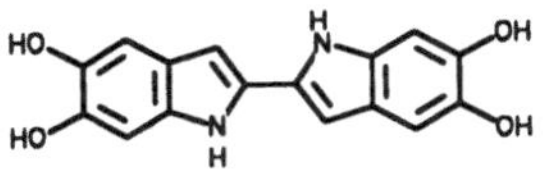

(31) 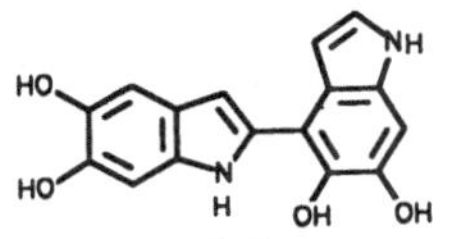   (32) 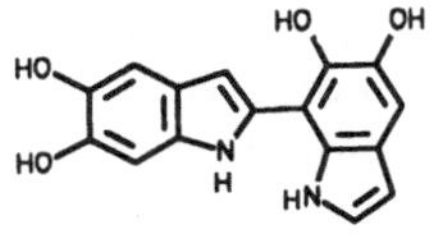   (33)

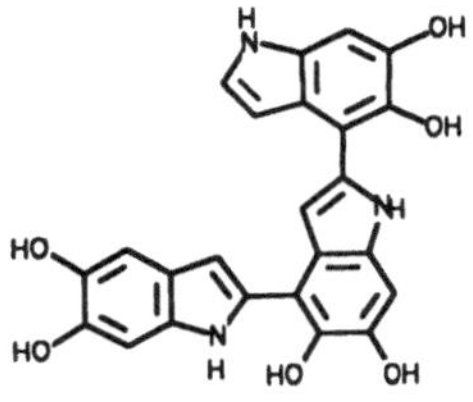

(34) 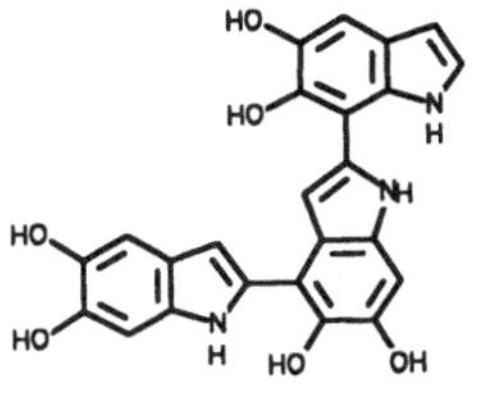   (35)

in aqueous buffer at pH 6.8. Unexpectedly, pigment formation was found to proceed slowly even in the presence of a large amount of enzyme. Moreover, when the reaction was stopped at the melanochrome stage, work-up of the oxidation mixture gave, besides much unchanged starting material, a 5–10% yield of another dimer which proved to be the 2,4-biindolyl (32).

In seeking for other oxidative enzymes which could possibly be involved in melanogenesis, it was found that peroxidase is far more effective than tyrosinase in catalyzing the oxidative polymerization of DHI (97, 98). Analysis of the peroxidase catalyzed reaction with the usual work-up procedure gave a well defined pattern of products consisting, besides the dimer (32), of three new oligomers which accounted for about 20% of the starting indole. These were isolated by preparative HPLC and identified as the 2,7-dimer (33) and the related trimers (34) and (35).

Overall, these results argue strongly against the previously suggested involvement of the 3-position in the building up of the eumelanin polymer(s). Indeed, from consideration of the structure of the oligomers so far identified, it appears that the dominant mode of coupling of DHI is via the 2- and 4-positions of the indole ring, with a minor contribution of the 7-position.

The mechanism of polymerization of DHI has not yet been adequately characterized; the polymerization is extremely rapid and may be accompanied by other reactions, for example peroxidative cleavage of the oligomers being formed. However, by analogy with other peroxidase-promoted oxidations, a realistic reaction pathway could involve an initial one electron transfer from DHI to the peroxidase-hydrogen peroxide complex to yield the DHI semiquinone (or an electronically equivalent species) to which a number of mechanistic options are offered, *i.e.* homolytic self-coupling or further one-electron oxidation and disproportionation. Subsequent attack of the nucleophilic 2-position of DHI to the electron-deficient 4- or 7-positions of the 5,6-indolequinone thus generated would then account for the observed mode of polymerization.

Fast reaction techniques, like flash photolysis and pulse radiolysis, have recently been applied to study reactive melanogenic intermediates which are too short lived to be detected with normal methods. A detailed account of these studies is given by LAND (*99*). Using a similar approach ADAMS and coworkers (*100, 101*) obtained evidence that the semiquinone radicals generated by one electron oxidation of DHI decay by first order kinetics to give a transient species with an absorption maximum at 430 nm. This is converted to more stable products exhibiting a maximum at 500 nm by first order kinetics with a half life of 10 ms

From a comparison of the kinetic and spectral features of the 430 nm species with those of the oxidation products of the model compounds 6-hydroxy-5-methoxyindole and 5-hydroxy-6-methoxyindole it was inferred that the dominating species formed by DHI oxidation is the quinone methide (**36**) which is in equilibrium with the quinoneimine (**37**) and the corresponding quinone (**38**). Similar results have been reported by LAMBERT *et al.* (*102*). Significantly, they found that the decay of the 430 nm species was halved when the oxidation was performed in D_2O rather than in H_2O. These and other observations were taken as evidence for the intermediacy of a trihydroxyindole, supposedly the 2,5,6-trihydroxyindole (**39**), arising by nucleophilic addition of water to the quinone methide. However, that a trihydroxyindole may be generated under biologically relevant conditions has not yet been demonstrated.

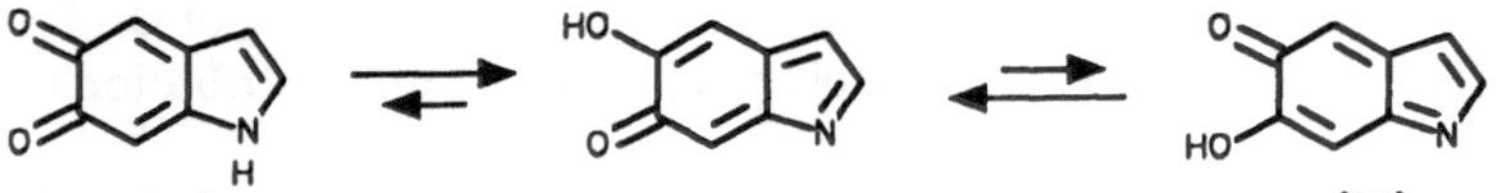

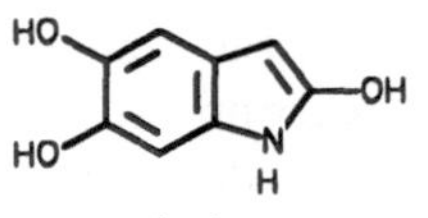

b. 5,6-Dihydroxyindole-2-Carboxylic Acid

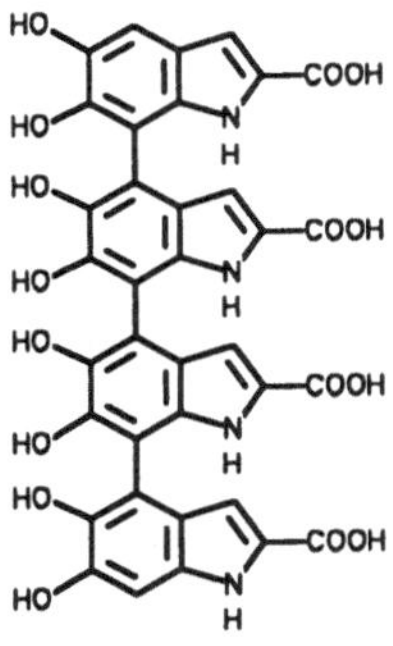

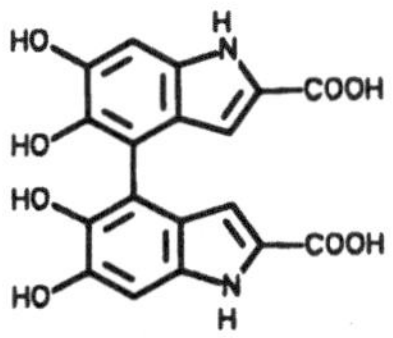

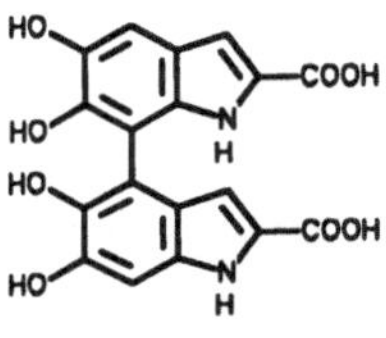

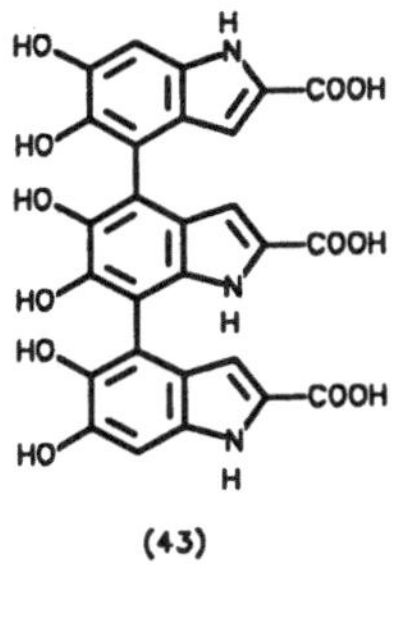

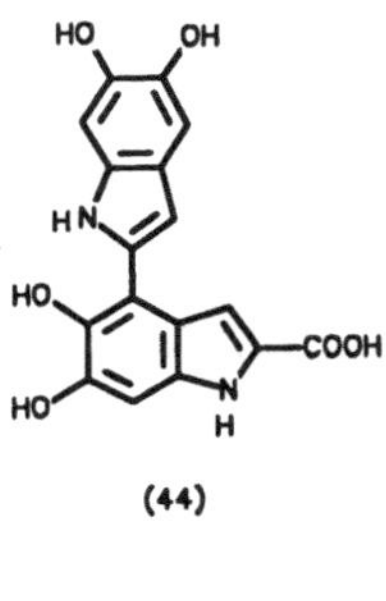

The oxidation chemistry of DHICA is in principle simpler than that of DHI, because the reactivity of the pyrrole ring is reduced by the presence of the carboxyl group. Accordingly, polymerization of DHICA must inevitably involve both the 4- and 7-positions, as first suggested by CROMARTIE and HARLEY-MASON (*92*). Some of the postulated oligomers, including a tetramer (**40**), were found by ITO and NICOL (*103*) in the *tapetum lucidum* of the catfish, and were considered to arise biogenetically via tyrosinase catalyzed oxidation of DHICA. When performed *in vitro*, this reaction gave an ill-defined pattern of oligomers, three of which could eventually be isolated (*104*) as the acetyl methyl ester derivatives and identified as the symmetrical dimer (**41**), the 4,7-dimer (**42**) and the trimer (**43**). As in the case of DHI, however, the kinetics of the reaction were rather slow. Accordingly, the oxidation with peroxidase was subsequently investigated (*98*). Under these latter conditions, an almost instantaneous reaction was observed which led in the early stages to a purple pigment with a maximum at 560 nm reminiscent of melanochrome. Reduction and acetylation of the oxidation mixture at this stage gave a distinct pattern of oligomers, consisting mainly of the 4,7-dimer (**42**) and the trimer (**43**) in about 10% and 25% yields, respectively. Given the occurrence of peroxidase in melanogenic compartments (*76*), these biomimetic experiments would suggest that this enzyme, rather than tyrosinase, is mainly involved in the oxidative conversion of DHI and DHICA to melanin.

In an extension of these studies (105), the peroxidase catalyzed oxidation of DHICA in the presence of DHI was also studied in order to appraise whether and to what extent copolymerization of the two melanogenic precursors can occur under biologically relevant conditions. Chemical analysis of the oxidation mixture as above revealed a complex but well-defined pattern of oligomers. The bulk of these consisted of DHI and DHICA oligomers, but one, isolated in 5% yield, proved to be the mixed dimer (44) arising presumably by trapping of 5,6-indolequinone-2-carboxylic acid by the nucleophilic 2-position of DHI. Consistent with this mechanism, the yield of formation of this dimer increased with increasing the DHI/DHICA molar ratio used in the cooxidation experiments. The isolation of the cross-coupling product provides the first evidence that DHI and DHICA can copolymerize under biologically relevant conditions. This would lead to a concept of eu-melanins as being intimate mixtures of homopolymers of DHI and DHICA, and copolymers of the two indoles, at different degrees of oxidation and polymerisation. The ratio of formation of DHI and DHICA from the rearrangement of dopachrome may affect the relative proportion of mixed-type oligomers vs. homopolymers.

V. Pheomelanins and Related Pigments

Unlike eumelanins, which are of widespread occurrence in the animal kingdom, authentic pheomelanins have been found so far only in epidermal tissues, such as skin and hair, where they usually occur together with the biogenetically related trichochromes (2). Rust-coloured pigment granules have been occasionally observed in the melanophores of some species of leaf frogs (Phyllomedusinae), but on analysis these were found to consist of pterorhodin (45), a red pteridine dimer (106). Another example of a presumptive pheomelanin is adenochrome which confers the characteristic purple colour to the branchial heart of certain cephalopods, e.g. Octopus vulgaris. Known since 1906 for its ability to chelate iron (III), the pigment was first isolated in a crude form by Fox and UPPDEGRAFF (107), who suggested a tentative formula $C_{41}H_{72}N_{12}O_{25}S_2$. Subsequent studies at Naples (108), however, showed that the pigment is in fact a complex mixture of unusual peptides derived from glycine and three novel aminoacids, adenochromine A (46), B (47) and C (48), which account for chelate formation with Fe(III). Biogenetically, the ad-

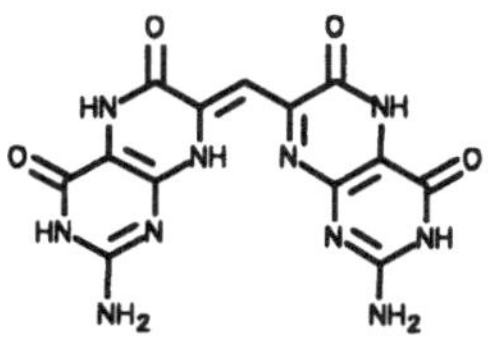

(45)

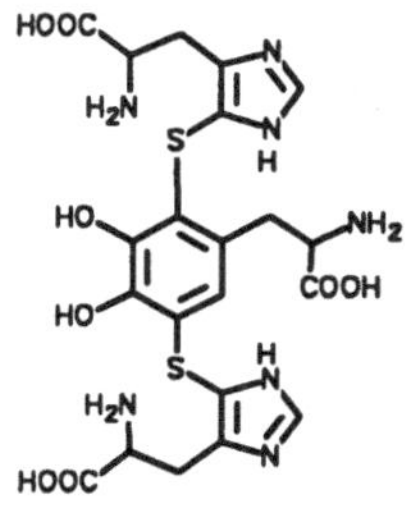

(46)

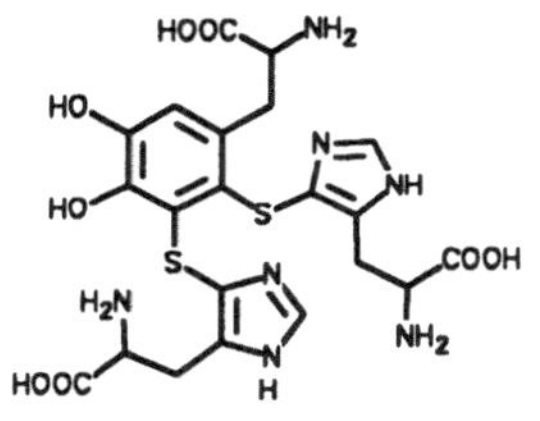

(47)

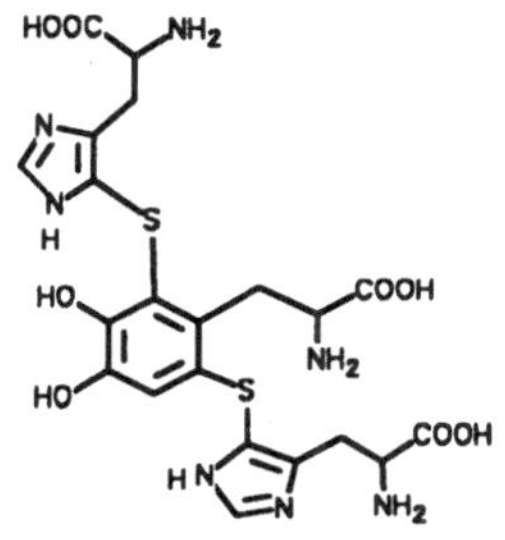

(48)

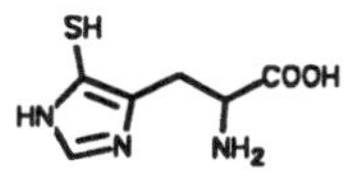

(49)

enochromines arise presumably by addition of the new amino acid 4-thiolhistidine (49) to dopaquinone. The analogy of such a pathway with the early stages of pheomelanin formation is remarkable (see later).

A. Trichochromes

The structures of the trichochromes so far identified from natural sources are shown in Fig. 12. The more frequently encountered trichochromes B (50) and C (51) were first isolated from the red feathers of New Hampshire hens and were later found in hair of red-headed people and of several mammals, including red-haired cattle, the golden hamster, New Zealand red rabbit, and red mongrel dog (3). Apparently, trichochromes E (54) and F (55) do not occur in red human hair and it is not clear whether the small amounts found in red feathers and in other sources arise as artifacts during the isolation procedure (109, 110). Notably, trichochromes B and C, but not E and F, are found in the urine of patients with melanoma metastasis (111).

The chemistry of the trichochromes has been repeatedly reviewed (2, 112, 113). A remarkable feature of these pigments is the bi-benzothiazine chromophore which contains a double conjugated system –S–C = C–C = N–, reminiscent of the polymethine cyanine dyes. In acid solution

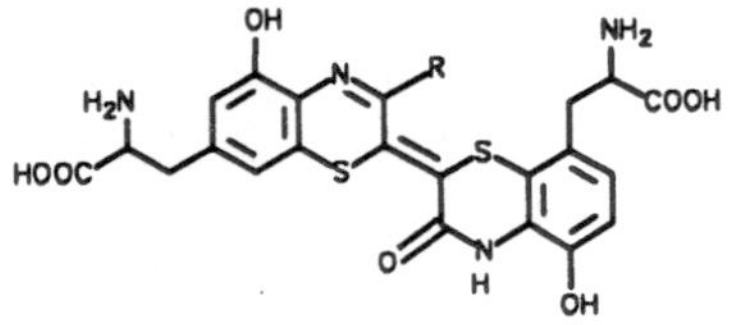

(50) R=COOH
(52) R=H

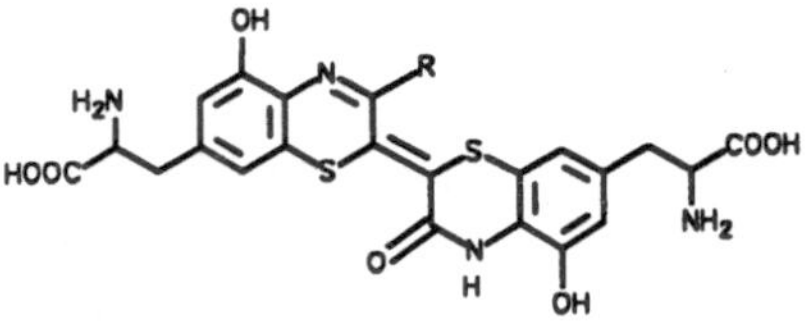

(51) R=COOH
(53) R=H

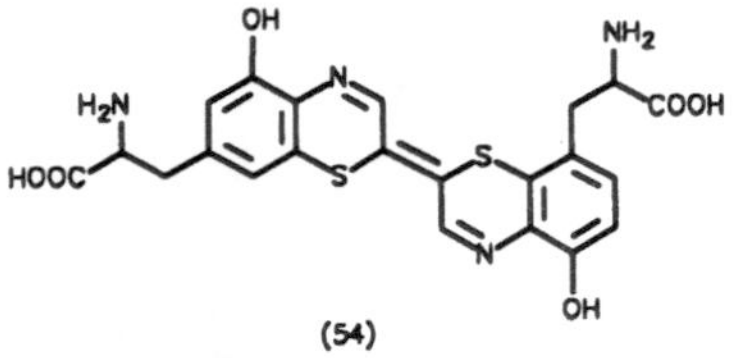

(54)

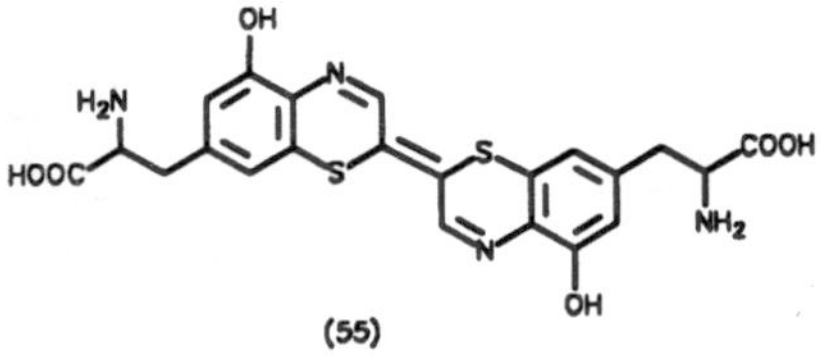

(55)

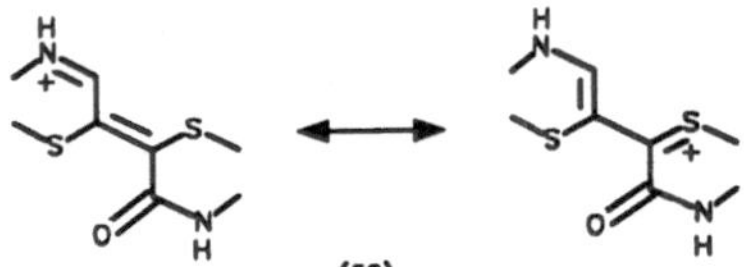

(56)

Fig. 12. Structure of trichochromes

protonation generates the mesomeric cation (**56**), resulting, in the case of trichochromes E and F, in a marked bathochromic shift of about 100 nm. These pigments therefore show pH-dependent colours, being deep violet in acid and red in alkaline solution. In trichochromes B and C, the presence of the carboxyl group at C-3 diminishes the basicity of the adjacent ring nitrogen, so that formation of the mesomeric cation does not normally occur and no acid shift of the absorption maximum is observed. However, on gentle heating in acid solution, they undergo selective decarboxylation at C-3 to give decarboxytrichochromes B (**52**) and C (**53**), which are yellow orange (λ 460 nm) in alkaline solution and red (λ 525 nm) in acids. This behaviour is useful for analysis, since extraction of hair or feathers with boiling mineral acids gives a mixture of indicator pigments which are more easily identified than the parent trichochromes.

An interesting reaction of trichochromes is the reductive cleavage of the chromophore with hydroiodic acid and red phosphorus. In the case of the symmetrically substituted trichochrome C, this reaction gives mainly the benzothiazinone amino acid (**57**) and 4-amino-3-hydroxyphenylalanine (**58**), along with smaller amounts of the benzothiazole aminoacid (**59**), arising evidently by ring contraction of the unsubstituted benzothiazine moiety (*114*). Under similar conditions, degradation of trichochrome B with hydroiodic acid and red phosphorus leads to the same benzothiazinone (**57**) and a mixture of the isomeric aminohydroxyphenylalanines (**58**) and (**60**). Thus, trichochromes B and C differ solely in the position of one alanine side chain on the $\Delta^{2,2'}$-bibenzothiazine ring system, which is fully consistent with biosynthetic studies (*115*).

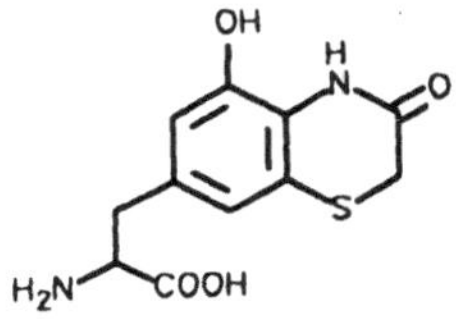

(57)

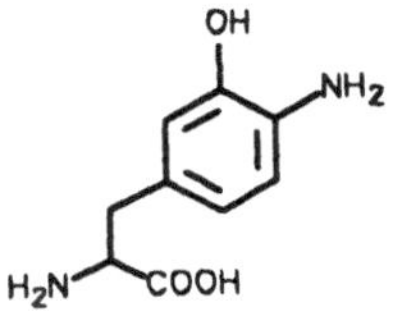

(58)

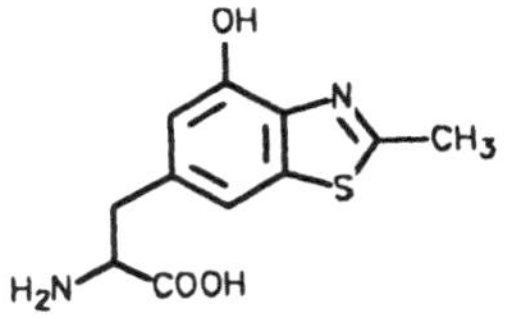

(59)

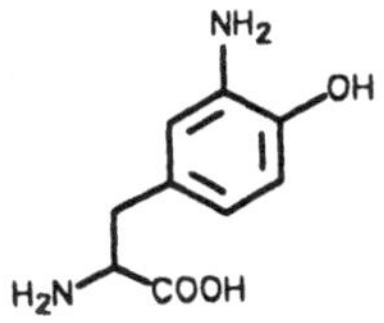

(60)

The basic chromophore of trichochromes B and C, *i.e.* (**63**), has been obtained by KAUL (*116*) by condensation of *o*-aminothiophenol (**61**) with mucochloric acid (**62**). A similar route has been utilized to obtain the

(61)

(62)

(63)

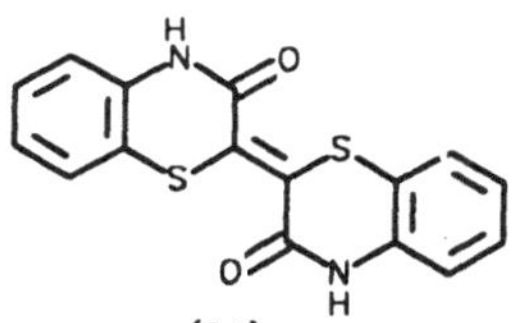

(64)

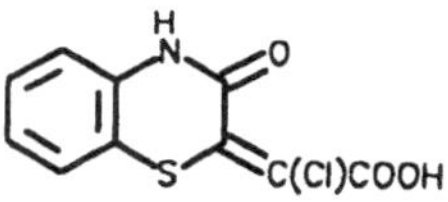

(65)

related benzothiazinone dye (**64**) by reaction of one mole of 2,3-dichloro-maleic anhydride with two moles of 2-aminothiophenol or its zinc salt in boiling acetic acid. This type of "inverse indigo synthesis" probably proceeds via the monocyclic adduct (**65**) which is the isolable product (74% yield) when equimolar quantities of the reagents are used.

As yet, the stereochemistry of the $\Delta^{2,2'}$-bibenzothiazine skeleton in the trichochromes has not been defined. Indirect evidence indicates that the isolated pigments may well be an equilibrium mixture of both geometrical isomers, the *trans* form predominating in the dark (*2*). This is supported (*117*) by a biomimetic-type synthesis of the parent ring system by aerobic oxidation of the 2H-1,4-benzothiazine (**67**), generated *in situ* by hydrolysis of 1-(*o*-aminophenylthio)-2,2-diethoxyethane (**66**). This proce-dure yields only the yellow *trans* isomer (**68a**) which in solution exhibits a pronounced photochromism, being converted into the red unstable *cis* (**68b**) form on brief irradiation with sunlight. In the dark the process reverses slowly at room temperature, but rapidly above 60 °C to regener-ate the *trans* isomer. Alternatively, the red isomer can be transformed into the *trans* isomer by protonation with dilute acids, to give the deep blue mesomeric cation (**69**) from which (**68a**) is obtained by neutralization (Fig. 13).

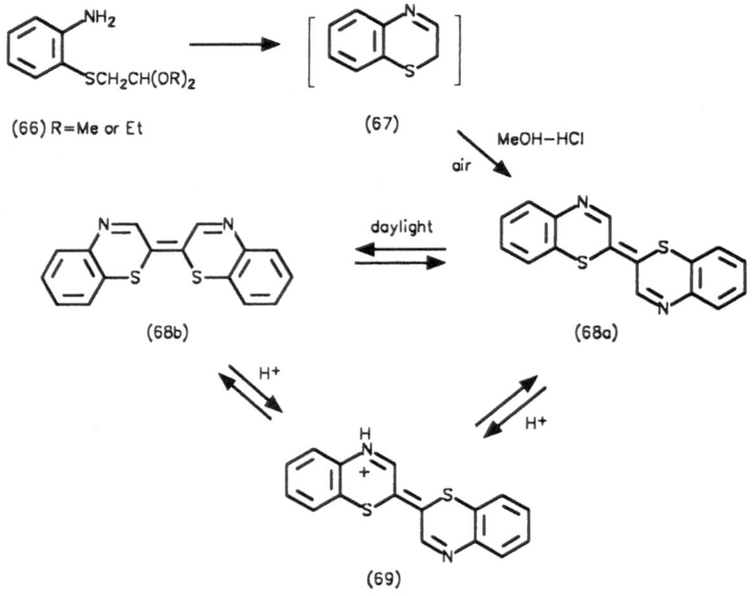

Fig. 13. Synthesis and interconversion of $\Delta^{2,2'}$-bi-(2H-1,4-benzothiazine)

B. Gallopheomelanins

1. Isolation and General Properties

Much of what is known about the chemistry of pheomelanins comes from studies of the pigments found in the feathers of New Hampshire hens (*Gallus gallus*). They were extracted with dilute alkali and separated into four main fractions, gallopheomelanins 1–4, by dialysis and chromatography on Sephadex or Biogel P columns (*118*). Another procedure developed more recently (*119*) involves preparative isoelectric focusing on a granulated bed. In any case, the gallopheomelanins thus obtained are inhomogeneous from the molecular viewpoint, each fraction consisting of mixtures of what are supposed to be polymers with similar physical and chemical characteristics.

Gallopheomelanin-1, the most abundant pigment, is protein-free, with an average molecular weight of about 2000 and a N/S ratio of 2.1, consistent with a close biogenetic link to the trichochromes (*118–120*). The other gallopheomelanin fractions (2–4) contain variable amounts of protein, up to 60%, and the possibility exists that they are chromoproteins. Attempts to characterize the protein-containing pigments by isoelectric focusing, apparent molecular weight, and aminoacid analysis have so far produced no definite answers (*121*). Little definite information is available on the redox state of gallopheomelanins, though they are generally described as being redox polymers (*112, 119, 121*). It is noteworthy, however, that gallopheomelanins and related pigments from red human hair exhibit free radical properties, due probably to the presence of a stable fraction of semiquinoneimines such as (**70**) within the pigment polymer (*122, 123*).

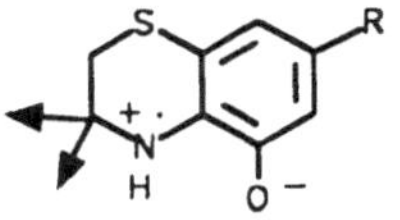

(70)

2. Gallopheomelanin-1

Like the trichochromes, gallopheomelanin-1 is zwitterionic in character and contains carboxyl, primary amino and phenolic groups. The absorption spectrum, however, shows no distinct chromophore, but a monotonic increase in the absorbance with decreasing wavelength and a barely detectable inflection between 290 and 320 nm. IR and NMR spectra also show broad bands of little diagnostic value. Thus, apart from the advantage of dealing with alkali soluble material, the structural problem is similar to that presented by the eumelanins.

On alkali fusion, the pigment gave a mixture of simple catechols and o-aminophenols, e.g. 4-amino-3-hydroxybenzoic acid (71) and 3-amino-4-hydroxybenzoic acid (72), while oxidation with hydrogen peroxide gave cysteic acid (73) in about 10% yield (118). Significantly, 5,6-dihydroxyindoles and pyrrolic acids, the typical products of eumelanin degradation, are not formed, an observation which implies the absence of 5,6-dihydroxyindole units in the pigment backbone. When heated in hydroiodic acid or hydrochloric acid, gallopheomelanin-1 gave (58) and the benzothiazole (74) with smaller amounts of the corresponding isomers (60) and (75) (118, 119). The synthesis of these new aminoacids has been reported by PATIL and CHEDEKEL (124). Other minor degradation

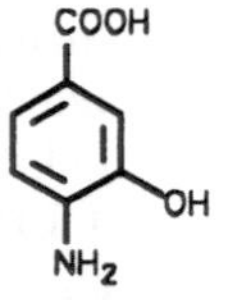

(71) 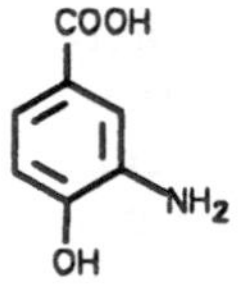   (72) 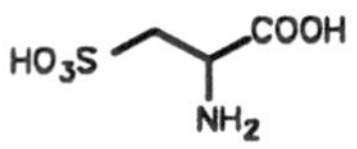   (73)

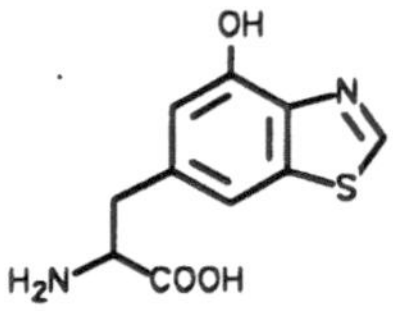

(74) 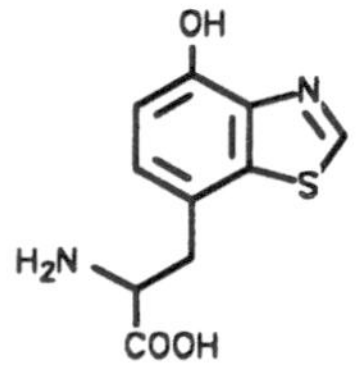   (75)

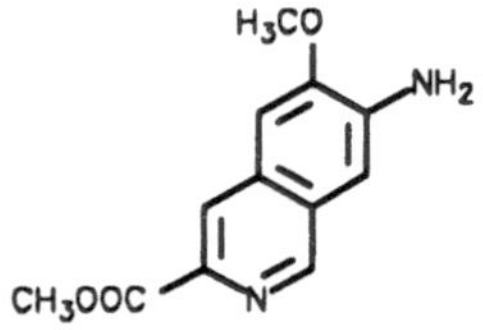

(76) 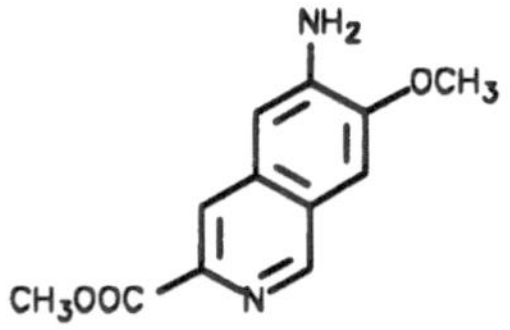   (77)

products of gallophaeomelanin-1 with HI include the benzothiazole amino acid (59) and the isoquinolines (76) and (77) isolated as the dimethylethers after treatment of the crude reaction mixture with diazomethane (125).

Some more insight into the structure of the pigment polymer resulted from oxidation of gallopheomelanin-1 with potassium permanganate which gave, besides the various thiazolic and pyridine acids, e.g. (78), (79) and (80), a larger fragment (0.1% yield) identified as the polyacid (81) (126–128). This was envisaged as arising by oxidative degradation of segments of the pigment backbone such as (82) containing tetrahydroiso-

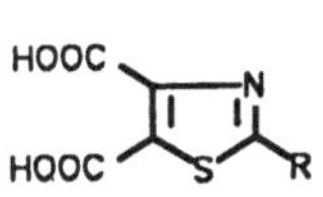

(78a) R=H
(78b) R=COOH
(78c) R=CONH₂

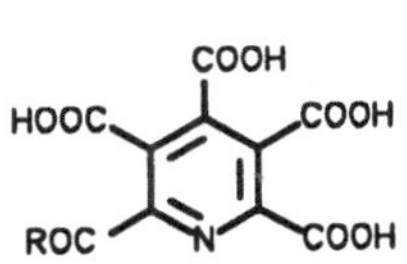

(79a) R=OH
(79b) R=NH₂

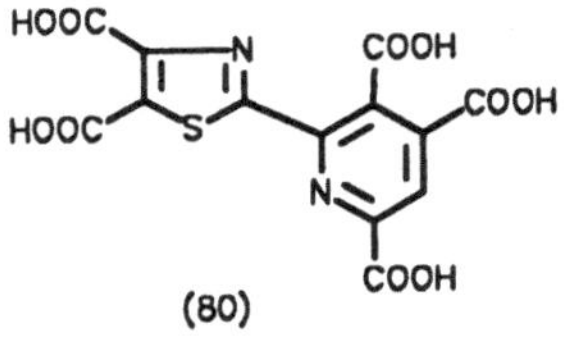

(80)

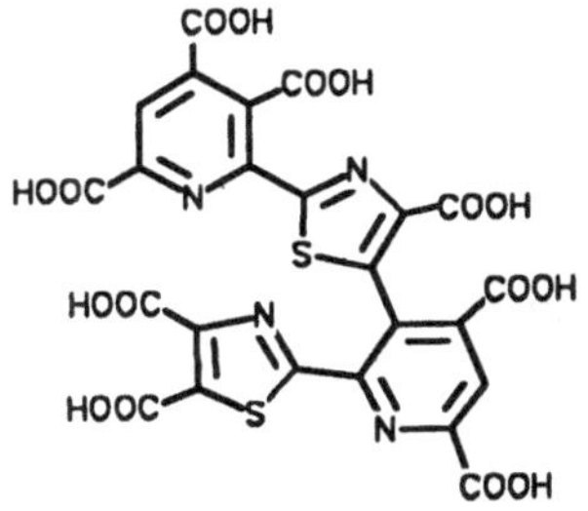

(81)

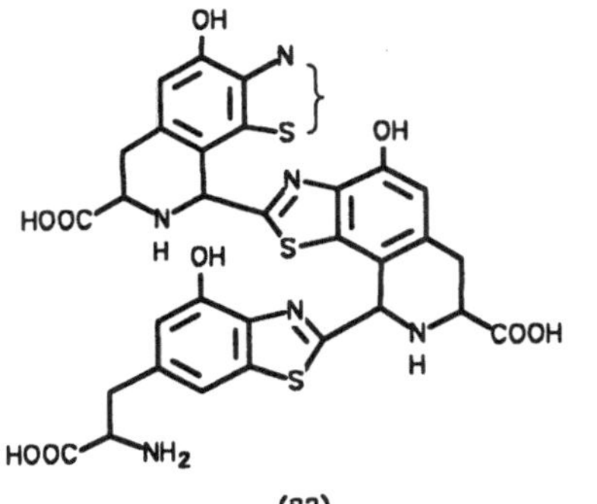

(82)

quinoline, benzothiazole and possible 1,4-benzothiazine units substituted at the 3-position.

Some of these structural units of the gallopheomelanin-1 molecules may be isomeric, e.g. (74) instead of (75), or at different level of oxidation, while others, not shown in (82), must contain uncyclized cysteine residues as indicated by formation of cysteic acid on oxidation with hydrogen peroxide. Other pheomelanins studied (129) gave a similar pattern of degradation products, suggesting that they have the same general structure but differ in the molecular size and relative proportion and/or sequence of various structural units in the polymeric chain.

While the presence of 1,4-benzothiazine units in the pigment polymers is generally accepted and consistent with biomimetic studies, the other concepts incorporated into the model structure, especially the benzothiazole and tetrahydroisoquinoline units, are questionable and probably need to be reconsidered (2). It is significant, for example, that some of the degradation products isolated from gallopheomelanins, e.g. the benzothiazole amino acids (59), (74) and (75) are also obtained by similar degradation of the trichochromes which, as we have seen, contain only 1,4-benzothiazinylalanine units. This would suggest that the benzothiazole fragments obtained from gallopheomelanins may be artifacts arising probably by acid-promoted ring contraction of similar benzothiazine units present in the pigment polymer. ITO and FUJITA (130) have also reported that permanganate oxidation of trichochrome F yields 4,5-thiazoledicarboxylic acid (78a) which is another characteristic fragment of gallopheomelanins. Against the proposed partial structure of pheomelanins are also the results of radiotracer studies indicating that the alanine side chain of dopa is predominantly incorporated intact into the pigment backbone (131, 132). It is clear, from these and other studies, that the problem of pheomelanin structure remains at present an open issue and would merit further investigation.

C. Biosynthetic Studies

As mentioned at the beginning of this chapter, all structural work on trichochromes and pheomelanins was guided by an early biomimetic experiment (27) in which dopa was enzymatically oxidized in the presence of cysteine. Under such conditions, the normal course of melanogenesis is deflected, yielding instead of the usual black precipitate a reddish brown pigment closely similar to the gallopheomelanins. Moreover, when the oxidation is stopped after a short time by acidification, the pale yellow solution turns to violet in a few minutes and a small but significant amount of trichochromes E and F is obtained. From these and other experiments, it was inferred that pheomelanins as well as trichochromes are formed *in vivo* by a deviation of the eumelanin pathway involving addition of cysteine (or a related sulphydryl compound) to nascent dopaquinone enzymatically generated by oxidation of dopa.

1. *The Pigment Precursors*

In model experiments using 4-methyl-1,2-benzoquinone (**83**), it was shown (*133*) that the initial reaction with cysteine involves addition of the SH group onto the *o*-quinone ring to give mainly the catechol aminoacid (**84**). The same reaction performed with nascent 4-methyl-1,2-benzoquinone proceeds similarly; in neither case does the amino group add or cyclize to give an amino-quinone.

(83)   (84)

The putative pigment precursor, *i.e.* 5-S-cysteinyldopa (5-cysdopa, **87**), was obtained (*134*) by addition of cysteine ethyl ester to the relatively stable N-acetyldopaquinone ethyl ester (**85**) followed by purification of the resultant adduct(s) as the acetyl derivative (**86**) (Fig. 14). Subsequent hydrolysis with acids afforded the expected aminoacid as the major product (48% yield) along with smaller amounts (ca. 5%) of another adduct identified as the 2-S-cysteinyldopa isomer (2-cysdopa, **88**) (*135*).

Both 2- and 5-cysdopas were detected in the early stages of the original enzymatic oxidation mixture of dopa and cysteine. In control experiments, it was found that cysdopas are not good substrates of tyrosinase; however, if a catalytic amount of dopa is added to the incubation mixture they are smoothly oxidized to give eventually reddish brown pigments known as 2-cysdopapheomelanin and 5-cysdopaspheomelanin, respectively (*135*). On analysis, these proved to be closely similar to the synthetic pheomelanin obtained by direct enzymatic oxidation of dopa and cysteine, except that they are more homogeneous from the molecular viewpoint, as confirmed by more recent studies (*131*).

In another relevant study (*136*), the formation of cysdopas was studied under biomimetic conditions by reaction of cysteine with dopaquinone

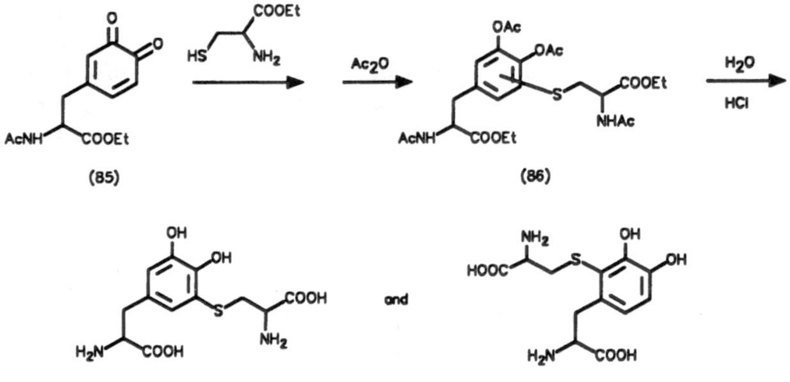

Fig. 14. Biomimetic synthesis of cysdopas

generated *in situ* by tyrosinase catalyzed oxidation of dopa. Fractionation of the reaction mixture by ion-exchange chromatography gave, in addition to 2- and 5-cysdopa, a small amount of the 6-isomer (**89**) as well as a diadduct identified as 2,5-dicystein-S,S-yldopa (dicysdopa, **90**). The yields of these products are in the order 14, 74, 1 and 5%, which is consistent with their relative contribution in the biosynthesis of the natural pigments. The involvement of all these intermediates in the biosynthesis of pheomelanins also accounts for the untractable nature and heterogeneous character of the pigment polymer.

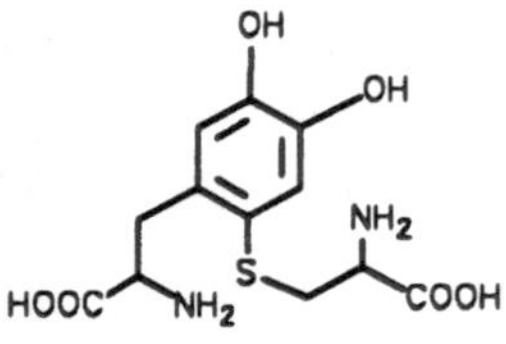

2. Occurrence and Origin of Cysdopas

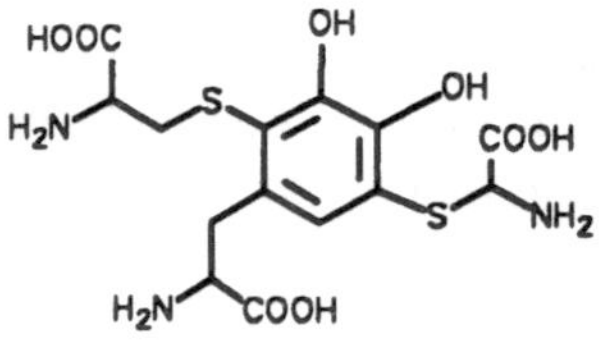

Another line of evidence for the close similarity between *in vitro* and *in vivo* pathways follows from parallel studies at Lund on the occurrence and distribution of cysdopas in man and other mammals. The major of these metabolites, 5-cysdopa, was first detected in melanoma tissues and in the urine from patients with metastatic pigmented melanoma (*137*). Later, it was found also in the urine, serum, and epidermal tissues (skin and hair) of healthy subjects (*138*). The other isomers of 5-cysdopa, namely, 2- and 6-cysdopa, were also found in melanoma urine as was the diadduct dicysdopa (*139, 140*). This latter had been previously isolated by ITO and NICOL (*141*) from the *tapetum lucidum* of the catfish. Space limitations preclude a description of the details of the biomedical implications of these findings. We will only say that the major cysdopa isomer is now currently used as a specific metabolic marker for the diagnosis and follow up of melanoma metastasis (*6*). Of particular interest from the chemical viewpoint is the demonstration (*142*) that the ratios of formation of the isomeric cysdopas in the urine are similar in control and melanoma patients and correspond to those determined in model experiments by incubation of dopa and cysteine with tyrosinase (*136*). Such a good correlation suggests that formation of cysdopas *in vivo* and *in vitro* proceeds similarly, by direct addition of cysteine to dopaquinone. However, an alternative pathway, which has also been considered, involves reaction of dopaquinone with GSH, followed by enzymatic hydrolysis of the corresponding adducts, *i.e.* the glutathionyldopas (GSHdopas) (Fig. 15).

The enzymes necessary for this transformation, γ-glutamyltransferase and peptidase, are of wide occurrence in biological systems (*143*) and have also been detected in malignant melanoma cells (*140*). Which of the two possible pathways is operative *in vivo* has not yet been assessed. Differences in GSH concentration in guinea pigs of different colours and the

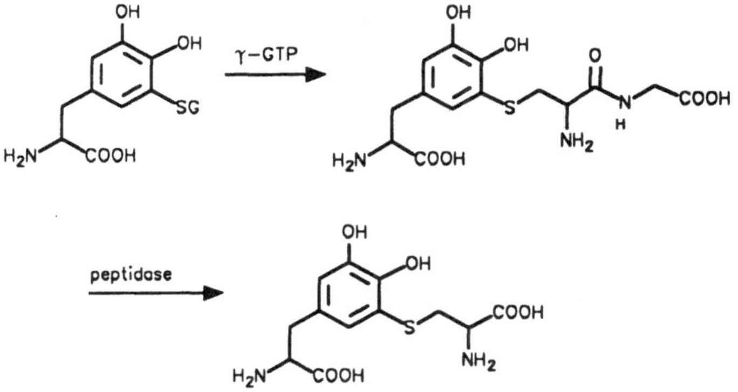

Fig. 15. Enzymatic conversion of 5-glutathionyldopa to 5-cysdopa

fact that GSH is by far the most abundant sulphydryl compound in cells point to an important role of GSHdopas in the biosynthesis of cysdopas (*144*). In support of this, ITO *et al.* (*145*) found that tyrosinase catalyzed co-oxidation of dopa and GSH proceeds in a similar fashion as with cysteine to give beside some 2,5-S,S-diglutathionyldopa (**94**) the three positional isomers 5-, 2-, and 6-GSHdopa (**91–93**) in comparable yields, *i.e.* 76, 12 and 5%, respectively.

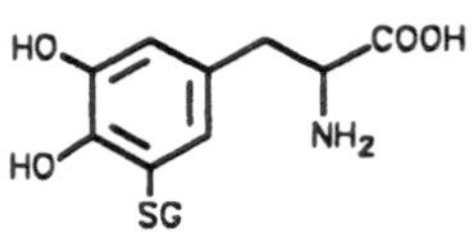

(91) 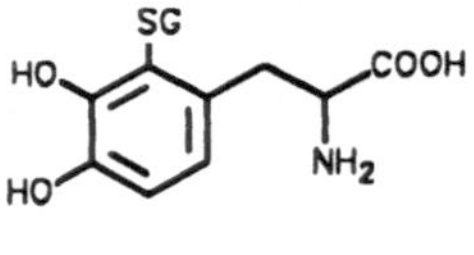   (92)

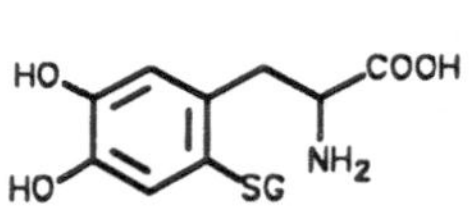

(93) 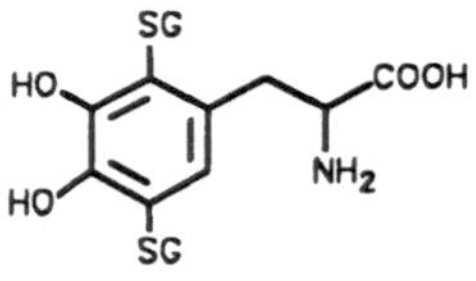   (94)

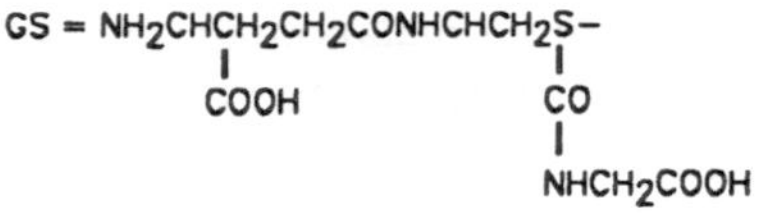

$$GS = \underset{\underset{COOH}{|}}{NH_2CHCH_2CH_2CONHCHCH_2S-}\quad\underset{\underset{NHCH_2COOH}{|}}{\underset{CO}{|}}$$

On the other hand, the fact that cysdopas but not GSHdopas are usually found in pigmented tissues would be in accord with the direct addition of cysteine to dopaquinone (*146*). As the lifetime of this quinone is very short, RORSMAN and co-workers suggested the existence in melanocytes of a compartment with higher concentrations of cysteine than GSH at the site of dopaquinone formation (*147*). However, evidence in support of this view comes mainly from studies on cultures melano-cytes, which may not reflect entirely the actual situation in epidermal tissues. The picture is further complicated by findings suggesting that formation of GSHdopas may also occur outside melanocytes (*148*) and that other oxidizing systems, *e.g.* peroxidase or superoxide generated radicals, may be involved as well (*149*).

3. 1,4-Benzothiazinylalanines

Figure 16 outlines the sequence of reactions which occur when 5-cysdopa is oxidized under biologically relevant conditions. It is of interest that the corresponding o-quinone (95) undergoes intramolecular cyclization of the amino group onto the carbonyl carbon to form the quinone imine (96). Analogous behaviour is displayed by the model adduct of cysteine to 4-methyl-1,2-benzoquinone (2, 150). Once formed, the quinoneimine can either give rise to the dihydrobenzothiazine (97) by redox exchange with the starting material or undergo rearrangement, with or without concomitant decarboxylation, to give (98) and (99).

Which of these products predominates depends on various parameters, including pH, oxygen tension, and notably the presence of metal ions. PALUMBO et al. (151), for example, found that zinc catalyzed autoxidation of 5-cysdopa leads to the benzothiazine acid (99) which corresponds to one of the structural units incorporated into trichochrome C. If the same reaction is performed in the absence of metal ions, the dihydrobenzothiazine acid (97) is formed in 72% yield (149).

As regards the conversion of these dihydrobenzothiazine derivatives into compounds of the trichochrome type, it is known that 2H-1,4-benzothiazines are rather prone to undergo self-coupling at the

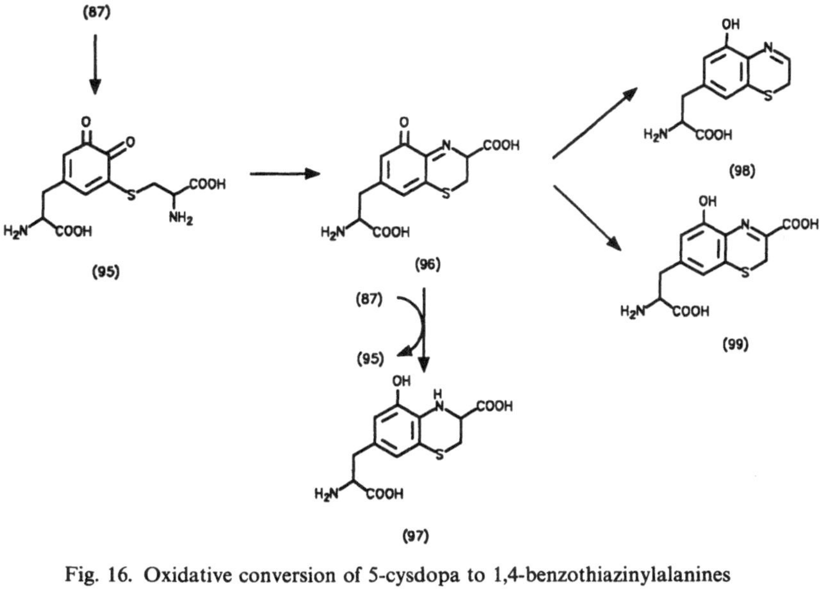

Fig. 16. Oxidative conversion of 5-cysdopa to 1,4-benzothiazinylalanines

2-position (*113*). In the case of the 3-phenyl derivative (**100**) formation of the dehydro dimer (**101**) proceeds merely by oxygenation in acidified ethanol at room temperature. Chloranil effects further dehydrogenation and both (**100**) and (**101**) give in good yield (**102**) in cold dioxane. Similarly (**103**) affords (**104**) in a reaction involving oxidative decarboxylation (*152*). The formation of a bibenzothiazine system in this way, presumably a radical process, is relevant for the biosynthesis of trichochromes, but the details of the reaction pathway *in vivo* remain elusive.

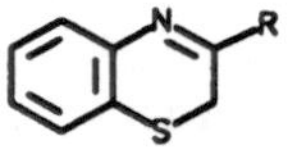

(100) R=Φ 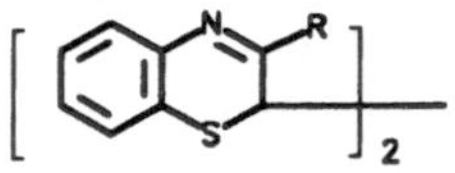   (101) R=Φ 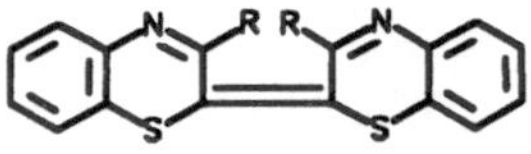   (102) R=Φ

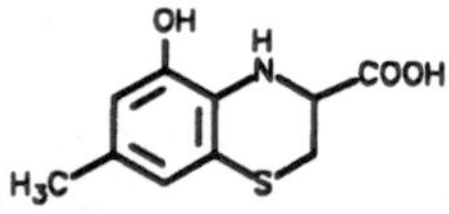

(103) 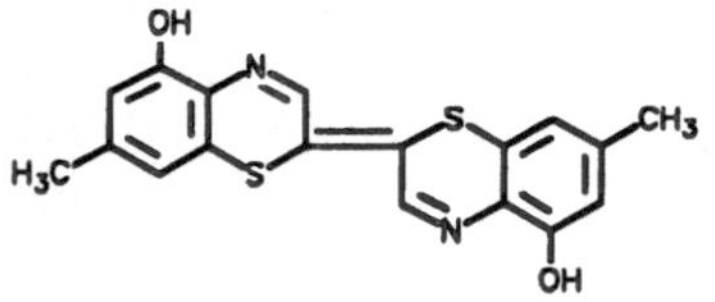   (104)

Little is also known about the later stages of the biosynthesis of pheomelanins after the formation of 1,4-benzothiazines. Using ^{13}C-, ^{14}C-, ^{35}S- and tritium-labelled cysdopa precursors, CHEDEKEL and co-workers (*131, 132*) presented evidence that during polymerization a small percentage of the benzylic carbons of the alanine side chain undergoes oxidation to an sp^2 olefinic or aromatic-type carbon, while the majority exhibit no change in oxidation state or attached functionality. Moreover, the ^{13}C-NMR spectrum of the pheomelanin pigment prepared by oxidation of 5-cysdopa labeled at the carboxyl group of the cysteine side chain indicates that the incorporated carbon is still in the form of carboxylic acid moieties.

More recently, PROTA and coworkers (*153*) reexamined the tyrosinase catalyzed oxidation of 5-cysdopa in phosphate buffer at pH 6.8. Periodic analysis of the mixture showed the rapid and almost quantitative conversion to the dihydrobenzothiazine (**97**), after which the reaction proceeds smoothly to give a discrete number of oligomeric products. Two of these, present in relatively larger amounts, were isolated and identified as two diastereoisomers corresponding to the gross structure (**105**). This could possibly arise by cycloaddition of the quinoneimine (**96**), formed by

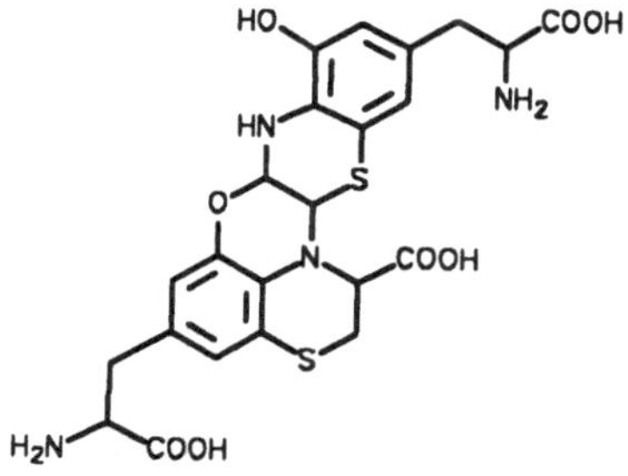

(105)

oxidation of the dihydrobenzothiazine (97), to the 4H-tautomer of the 1,4-benzothiazine (98), derived from decarboxylation of (96).

Compound (105) is the first product beyond the benzothiazine stage that has been isolated from the biomimetic oxidation of 5-cysdopa. Whether it plays a role in the biosynthesis of pheomelanins or merely represents a side branch of the pathway is still under assessment. It is noteworthy, however, that in acid solution (105) is unstable and rapidly gives rise to the violet trichochrome F (55), presumably via acid-catalyzed ring-opening and subsequent oxidative coupling of the resultant 1,4-benzothiazine (98) at the 2-position. The formation of trichochrome F (55) in this way is of interest in relation to the possibility that the pigment obtained from red hair and feathers (as well as the analogous E) may be artifactually derived from (105) (or the analogous dimer from 2-cysdopa) during the isolation procedure.

VI. Overview of Reaction Pathways

An overall view of the reaction pathways leading to eumelanins, pheomelanins and related metabolites is given in Figure 17. The amounts and types of the products formed in melanocytes are known to involve the interaction of multiple genes, some of which have recently been cloned and identified (4, 5). Among these, the albino locus gene holds a central position since it encodes the enzyme tyrosinase which catalyzes the initial and rate determining steps of melanogenesis, namely, the hydroxylation of tyrosine to dopa and its subsequent conversion to dopaquinone. The subsequent metabolic fate of this intermediate is the result of its intrinsic high reactivity coupled with the biochemical environment within melanocytes.

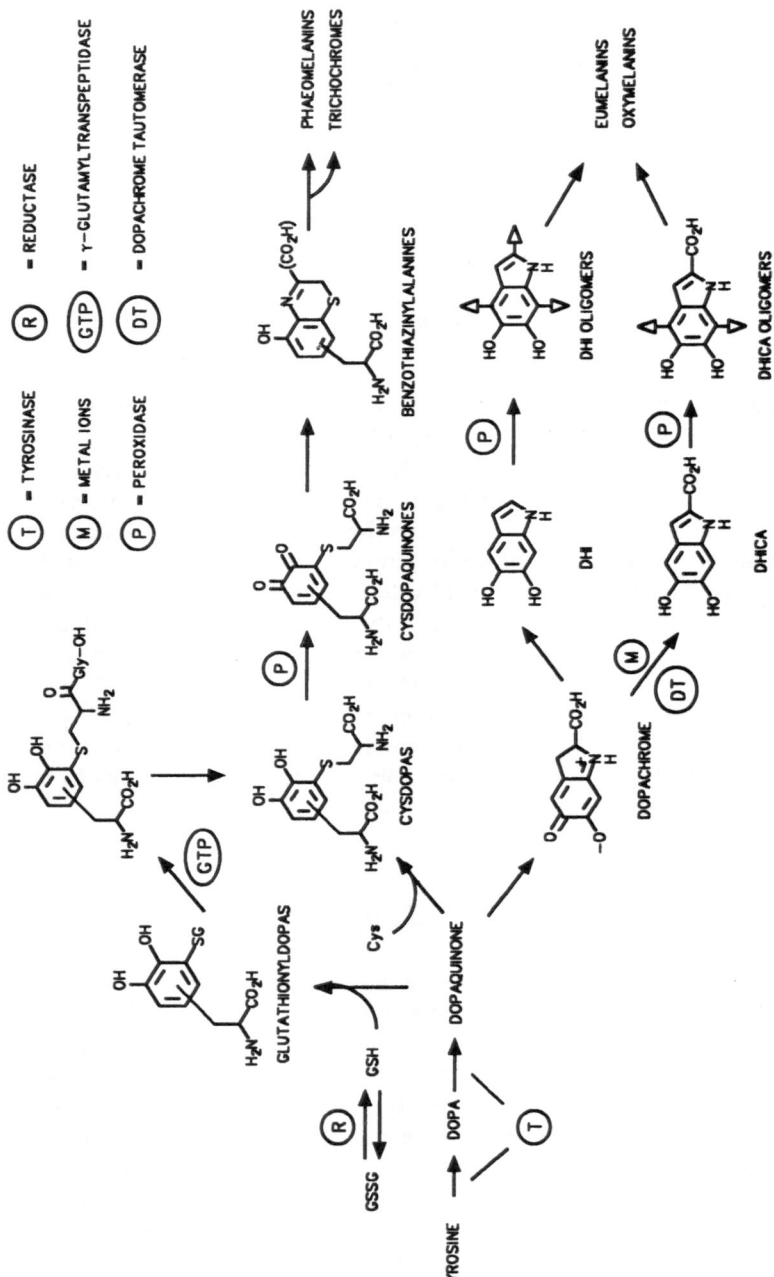

Fig. 17. Reaction pathways leading to melanins and related metabolites

In eumelanin-forming melanocytes, most of the dopaquinone is converted to leucodopachrome which is rapidly oxidized to dopachrome by a redox exchange reaction with dopaquinone itself, as correctly indicated by RAPER (12). As a result, a certain amount of dopa can be expected to be constantly present in active melanocytes in spite of the fact that tyrosinase can oxidize dopa more rapidly than tyrosine. The proposed sequence of reactions leading from tyrosine to dopachrome has been confirmed by various groups using electroanalytical techniques (154), pulse radiolysis (99) and EPR (155). Rate constants for the implicated chemical steps at different pH and temperature values have been calculated as well as the activation thermodynamic parameters for the deprotonation of dopaquinone and its subsequent cyclization to leucodopachrome (156, 157).

As pointed out earlier, there is now considerable evidence that within melanocytes dopachrome undergoes enzymatic or metal-catalyzed rearrangement to form DHI and DHICA. The subsequent polymerization of these indoles to eumelanins is probably associated with an oxidative stress leading to an increase in the intracellular level of reactive oxygen species. As a result, some hydrogen peroxide is eventually generated which, in combination with metal ions and/or peroxidase, can efficiently bring about the oxidative polymerization of both DHI and DHICA. Hydrogen peroxide is also responsible for the subsequent oxidative breakdown of part of the 5,6-dihydroxyindole units in the pigment polymers, accounting for the molecular heterogeneity of natural eumelanins. A more extensive attack of the pigment polymer by hydrogen peroxide could reasonably explain the formation of the oxymelanins, but this remains speculative at present.

Switch of the pathway to pheomelanins and trichochromes appears to be related primarily to an increased availability of sulphydryl compounds in the melanogenic compartment of melanocytes (144). In support of this mechanism is the now classical observation by CLEFFMAN (158) that in tissue culture pheomelanin-forming melanocytes from agouti mouse follicles produce only black pigment, but revert to pheomelanin production when sufficient GSH or cysteine is added to the nutrient medium. While the effect of cysteine on the synthesizing activity of melanocytes is consistent with the trapping of dopaquinone to form the cysdopas, the involvement of GSH in the biosynthesis of pheomelanins is less straightforward, since the resulting adducts to dopaquinone, the GSHdopas, need to be enzymatically converted to cysdopas to enter the pheomelanin pathway. This implies that, in addition to serving as a possible substrate for the synthesis of cysdopas, GSH could convert part of the dopaquinone formed in melanocytes into the non-melanogenic

GSHdopas. Hence, in the absence of the necessary enzymatic steps to cleave the glutamyl and glycyl residues, the formation of GSHdopas would only sidetrack dopaquinone without producing either pheomelanin or eumelanin.

Such a fine regulatory role of GSH provides a reasonable biochemical explanation for the understanding of qualitative differences in the basal level of melanin pigmentation as seen for example in black and white skin (7). A similar mechanism has also been implicated to account for the enhanced pigment synthesis associated with certain chronic or acute conditions, such as inflammation and UV-induced erythema, which are known to deplete the cutaneous GSH levels.

Of course, a number of other regulatory factors may still be missing from this general picture. However the basic chemistry of melanogenesis appears, at least for the present, to be established and so provides the conceptual framework for a molecular approach to the control of epidermal pigmentation and its application to human diseases.

Abbreviations

Cysdopas, cysteinyldopas; DHI, 5,6-dihydroxyindole; DHICA, 5,6-dihydroxy-indole-2-carboxylic acid; dicysdopa, 2,5-dicystein-S,S-yldopa; GSH, glutathione; GSHdopas, glutathionyldopas; PDCA, pyrrole-2,3-dicarboxylic acid; PTCA, pyrrole-2,3,5-tricarboxylic acid.

References

1. SWAN, G.A.: Structure, Chemistry and Biosynthesis of the Melanins. Fortschr. Chem. Organ. Naturstoffe, 31, 522 (1974).
2. PROTA, G.: Progress in the Chemistry of Melanins and Related Metabolites. Med. Res. Rev., 8, 525 (1988).
3. PROTA, G.: Melanins and Melanogenesis. San Diego: Academic Press. 1992.
4. HEARING, V.J., and K. TSUKAMOTO: Enzymatic Control of Pigmentation in Mammals. FASEB J., 5, 2902 (1991).
5. TAKEUCHI, T.: Molecular Structure of the Tyrosinase Gene. In: The Pigment Cell: From the Molecular to the clinical Level (Y. MISHIMA, ed.). Copenhagen: Munksgaard. 1992.
6. RORSMAN, H., and S. PAVEL: Metabolic Markers and Melanoma. In: Cutaneous Melanoma Biology and Management (N. CASCINELLI, M. SANTINAMI and U. VERONESI, eds.), p. 79. Milano: Masson. 1990.
7. MONTAGNA, W., G. PROTA, and J. KENNEY: Black Skin. New York: Academic Press. 1993.
8. LE CAT, C.N.: TraitÅ de la Couleur de la Peau Humaine en general, de celle des Nègres en particulier, et de la Metamorphose d'une de ces Couleurs en l'autre, soit de Naissance, Soit accidentellement. Amsterdam: M. M. Rey. 1765.

9. BILLINGHAM, R.E., and W.K. SILVERS: The Melanocytes of Mammals. Q. Rev. Biol., 35, 1 (1960).
10. BERTRAND, G.: Sur une nouvelle oxydase, ou ferment soluble oxydant, d'origine vegetale. Comp. Rend. Acad. Sci. (Paris), 122, 1215 (1996).
11. ROBB, D.A.: Tyrosinase. In: Copper Proteins, Vol. 2 (R. LOUTIE, ed.), p. 207. Boca Raton: CRC Press 1984.
12. RAPER, H.S.: The Aerobic Oxidases. Physiol. Rev., 8, 245 (1928).
13. RAPER, H.S.: Some Problems of Tyrosine Metabolism. J. Chem. Soc., 125 (1938).
14. BEER, R.J.S., K. CLARKE, H.G. KHORANA, and A. ROBERTSON: The Chemistry of the Melanins, Part I: The Synthesis of 5,6-Dihydroxyindole and Related Compounds. J. Chem. Soc., 2223 (1948).
15. BEER, R.J.S., T. BROADHURST, and A. ROBERTSON: The Chemistry of the Melanins, Part V: The Autoxidation of 5,6-Dihydroxyindoles. J. Chem. Soc., 1947 (1954).
16. MASON, H.S.: The Structure of Melanin. In: The Pigmentary System: Advances in Biology of Skin, Vol. 8 (W. MONTAGNA and F. HU, eds.), p. 293. Oxford: Pergamon Press. 1967.
17. NICOLAUS, R.A.: Melanins. Paris: Hermann. 1968.
18. BLOIS, M.S.: The Melanins: Their Synthesis and Structure. Photochem. Photobiol. Rev., 3, 115 (1978).
19. QUEVEDO Jr., W.C., T.B. FITZPATRICK, G. SZABO, and K. JIMBOW: Biology of the Melanin Pigmentary System. In: Dermatology in General Medicine, Vol. 1 (T.B. FITZPATRICK, H.S. EISEN, K. WOLFF, M. FREEDBERG, and K.F. AUSTEN, eds.), p. 224. New York: McGraw-Hill. 1987.
20. JIMBOW, K., T.B. FITZPATRICK, and M.M. WICK: Biochemistry and Physiology of Melanin Pigmentation. In: Biochemistry and Physiology of Skin, Vol. 2 (L.A. GOLD-SMITH, ed.), p. 873. Oxford: Pergamon Press. 1991.
21. FITZPATRICK, T.B., P.C.J. BRUNET, and A. KUKITA: The Nature of Hair Pigment. In: The Biology of Hair Growth (W. MONTAGNA, ed.), p. 225. New York: Academic Press. 1958.
22. SORBY, H.C.: On the Colouring Matters Found in Human Hair. J. Anthropol. Inst. Lond., 8, 1 (1878).
23. FLESCH, P., and S. ROTHMAN: Isolation of an Iron Pigment from Human Red Hair. J. Invest. Dermatol., 6, 257 (1945).
24. BARNICOT, N.A.: The Pigment, Trichosiderin, from Human Red Hair. Nature, 77, 528 (1956).
25. BOLDT, P.: Zur Kenntnis des Trichosiderins eines Pigments aus roten Haaren. Naturwissenschaften, 51, 265 (1964).
26. BOLDT, P., and E. HERMESTEDT: Pyrrotrichole, eine Gruppe farbiger Verbindungen aus rotem Menschenhaar. Z. Naturforsch., 22B, 718 (1967).
27. PROTA, G., and R.A. NICOLAUS: Struttura e biogenesi delle feomelanine. Nota I: Isolamento e proprieta' dei pigmenti delle piume. Gazz. Chim. Ital., 97, 666 (1967).
28. PROTA, G., and R.H. THOMSON: Melanin Pigmentation in Mammals. Endeavour, 35, 31 (1976).
29. PROTA, G.: Melanin and Pigmentation. In: Coenzymes and Cofactors, Vol. 3 (D. DOLPHIN, R. PAULSON, and O. ABRAMOVIC, eds.), p. 441. New York: Wiley & Sons. 1989.
30. WHEELER, M.H., and A.A. BELL: Melanins and Their Importance in Pathogenic Fungi. Curr. Top. Med. Mycol., 7, 338 (1988).
31. PETER, M.G.: Chemical Modification of Biopolymers by Quinones and Quinones Methides. Angew. Chemie Int. Ed. Engl., 28, 555 (1989).

32. PIATTELLI, M., E. FATTORUSSO, S. MAGNO, and R.A. NICOLAUS: The Structure of Melanins and Melanogenesis, III: The Structure of Sepiomelanin. Tetrahedron, 19, 2061 (1963).

33. ITO, S., Y. IMAI, K. JIMBOW, and K. FUJITA: Incorporation of Sulfhydryl Compounds into Melanins in vitro. Biochim. Biophys. Acta, 964, 1 (1988).

34. CARSTAM, R., C. HANSSON, C. LINDBLADH, H. RORSMAN, and E. ROSENGREN: Dopaquinone Addition Products in Cultured Human Melanoma Cells. Acta Derm. Venereol. (Stockh.), 67, 100 (1987).

35. D'ISCHIA, M., A. NAPOLITANO, and G. PROTA: Sulphydryl Compounds in Melanogenesis, Part I: Reaction of Cysteine and Glutathione with 5,6-Dihydroxyindoles. Tetrahedron, 43, 5351 (1987).

36. D'ISCHIA, M., A. NAPOLITANO, and G. PROTA: Sulphydryl Compounds in Melanogenesis, Part II: Reactions of Cysteine and Glutathione with Dopachrome. Tetrahedron, 43, 5357 (1987).

37. BU'LOCK, J.D., and J. HARLEY-MASON: Melanin and Its Precursors, Part II: Model Experiments on the Reactions between Quinones and Indole, and Consideration of a Possible Structure for the Melanin Polymer. J. Chem. Soc., 703 (1951).

38. YASUNOBU, K.T., E.W. PETERSON, and H.S. MASON: The Oxidation of Tyrosine-Containing Peptides by Tyrosinase. J. Biol. Chem., 234, 3291 (1959).

39. ROSEI, M.A., L. MOSCA, and C. DE MARCO: Melanin Production from Enkephalins by Tyrosinase. Biochem. Biophys. Res. Commun., 184, 1190 (1992).

40. BENATHAN, M., and H. WYLER: Contribution a l'analyse quantitative des melanines. Yale J. Biol. Med., 53, 389 (1980). For details see: BENATHAN, M.: Ph.D. Thesis, Lausanne University. 1980.

41. CRIPPA, P.R., V. HORAK, G. PROTA, P. SVORONOS, and L. WOLFRAM: Chemistry of Melanins. In: The Alkaloids, Vol. 36 (A. BROSSI, ed.), p. 253. New York: Academic Press. 1989.

42. ITO, S.: Reexamination of the Structure of Eumelanin. Biochim. Biophys. Acta, 883, 155 (1986).

43. CLARK, M.B.J., J.A.J. GARDELLA, T.M. SCHULTZ, D.G. PATIL, and L.J. SALVATI: Solid-State Analysis of Eumelanin Biopolymers by Electron Spectroscopy for Chemical Analysis. Anal. Chem., 62, 949 (1990).

44. AIME, S., M. FASANO, and C. CROOMBRIDGE: Solid-State Carbon-13 NMR Characterization of Melanin Free Acids from Biosynthetic and Natural Melanins. Gazz. Chim. Ital., 120, 663 (1990).

45. CHEDEKEL, M.R., D.G. PATIL, K.V. RAO, B.P. MURPHY, M. CLARK, J. GARDELLA, and T.M. SCHULTZ: Solid Phase Carbon-13 NMR of ^{13}C-Enriched Eumelanins: The Fate of the Pyrrolic Ring. Pigment Cell Res., 1, 282 (1988) (Abs.).

46. SWAN, G.A., and A. WAGGOTT: Studies Related to the Chemistry of Melanins, Part X: Quantitative Assessment of Different Types of Units Present in Dopa-Melanin. J. Chem. Soc. (C), 1409 (1970).

47. BU'LOCK, J.D.: The formation of Melanin from Adrenochrome. J. Chem. Soc., 52 (1961).

48. CORRADINI, M.G., O. CRESCENZI, and G. PROTA: A Reinvestigation of the Anaerobic Conversion of Adrenochrome to "Adrenaline Black". Tetrahedron, 44, 1803 (1988).

49. SEALY, R.C., C.C. FELIX, J.S. HYDE, and H.M. SWARTZ: Structure and Reactivity of Melanins: Influence of Free Radicals and Metal Ions. In: Free Radicals in Biology, Vol. 4 (W.A. PRYOR, ed.), p. 209. New York: Academic Press. 1980.

50. SARNA, T.: Properties and Function of the Ocular Melanin – a Photobiophysical View. J. Photochem. Photobiol. B: Biol (1991).

51. CHEDEKEL, M., A.B. AHENE, and L. ZEISE: Melanin Standard Method: Empirical Formula 2. Pigment Cell Res., 5, 240 (1992).

52. GALVAO, D.S., and M.J. CALDAS: Theoretical Investigation of Model Polymers for Eumelanins, II: Isolated Defects. J. Chem. Phys., 93, 2848 (1990).

53. GALVAO, D. S., and M.J. CALDAS: Theoretical Investigation of Model Polymers for Eumelanins, I: Finite and Infinite Polymers. J. Chem. Phys., 92, 2630 (1990).

54. McGINNISS, J.E., P. CORRY, and P. PROCTOR: Amorphous Semiconductor Switching in Melanins. Science, 183, 853 (1974).

55. SARNA, T., and H.M. SWARTZ: Interaction of Melanin with Oxygen (and Related Species). In: Atmospheric Oxidation and Antioxidants, Vol 3 (G. SCOTT, ed.). Amsterdam: Elsevier. 1991.

56. KORYTOWSKI, W., B. PILAS, T. SARNA, and B. KALYANARAMAN: Photoinduced Generation of Superoxide Anion and Hydroxyl Radical in Melanins. Photochem. Photobiol., 45, 185 (1987).

57. NICOLAUS, R.A., and M. PIATTELLI: Structure of Melanins and Melanogenesis. J. Pol. Sci., 58, 1133 (1962).

58. NICOLAUS, R.A., M. PIATTELLI, and E. FATTORUSSO: The Structure of Melanins and Melanogenesis, IV: On Some Natural Melanins. Tetrahedron, 20, 1163 (1964).

59. ZEISE, L., and M.R. CHEDEKEL: Melanin Standard Method: Titrimetric Analysis. Pigment Cell Res., 5, 230 (1992).

60. CRESCENZI, O., M. D'ISCHIA, and G. PROTA: The Alleged Stability of Dopamelanin Revisited. Gazz. Chim. Ital. (in press).

61. PIATTELLI, M., E. FATTORUSSO, G.S. MANO, and R.A. NICOLAUS: The Structure of Melanins and Melanogenesis, II: Sepiomelanin and Synthetic Pigments. Tetrahedron, 18, 941 (1962).

62. NICOLAUS, R.A.: Melanins. In: Methodicum Chimicum, Vol. 11 (F. KORTE and M. GOTO, eds.), p. 190. Georg Thieme, Stuttgart, 1978.

63. BINNS, F., R.F. CHAPMAN, N.C. ROBSON, G.A. SWAN, and A. WAGGOTT: Studies Related to the Chemistry of Melanins, Part VIII: The Pyrrolecarboxylic Acids formed by Oxidation or Hydrolysis of Melanins Derived from 3,4-Dihydroxyphenylethylamine or DOPA. J. Chem. Soc. (C), 1128 (1970).

64. D'ISCHIA, M., A. PALUMBO, and G. PROTA: 5,6-Dihydroxyindole-2-carboxylic Acid by Treatment of Sepiomelanin with Sodium Borohydride. Tetrahedron Lett., 26, 2801 (1985).

65. REMERS, W.A.: Properties and Reactions of Indoles. In: Indoles, Part 1 (W.J. HOULIHAN, ed.), p. 152. New York: J. Wiley. 1972.

66. CHENG, A.C., A.T. SHULGIN, and N. CASTAGNOLI Jr.: Studies on the Chemical Reactivity on the Quinone Methide Derived from the Oxidative Cyclization of α-Methyl-3,4-dihydroxyphenylalanine Ethyl Ester. J. Org. Chem., 47, 5258 (1985).

67. CRESCENZI, O., C. COSTANTINI, and G. PROTA: Evidence for the Intermediacy of Quinone-Methides in the Rearrangement of Aminochromes to 5,6-Dihydroxyindoles. Tetrahedron Lett., 42, 6095 (1990).

68. SUGUMARAN, M., and V. SEMENSI: Formation of a Stable Quinone Methide During Tyrosinase-Catalyzed Oxidation of Alpha-Methyldopa Methyl Ester and Its Implication in Melanin Biosynthesis. Bioorg. Chem., 18, 144 (1990).

69. COSTANTINI, C., O. CRESCENZI, and G. PROTA: Mechanism of the Rearrangement of Dopachrome to 5,6-Dihydroxyindole. Tetrahedron Lett., 31, 3849 (1991).

70. PAWELEK, J.M.: After Dopachrome? Pigment Cell Res., 4, 53 (1991).

71. HEARING, V.J., T.M. EKEL, P.M. MONTAGUE, and J.M. NICHOLSON: Mammalian Tyrosinase. Stoichiometry and Measurement of Reaction Products. Biochim. Biophys. Acta, 611, 251 (1980).

72. STRAVS-MOMBELLI, L., and H. WYLER: Reinvestigation of the Formation of Dopa-Melanin. New Aspects of the Autoxidation of Dopa. In: Pigment Cell Biological Molecular and Clinical Aspects of Pigmentation (J. BAGNARA, S.N. KLAUS, E. PAUL, M. SCHARTL, eds.), p. 69. Tokyo: University of Tokyo Press. 1985.

73. BOWNESS, J.M., R. MORTON, M.H. SHAKIR, and A.L. STUBBS: Distribution of Copper and Zinc in Mammalian Eyes. Occurrence of Metals in Melanin Fractions from Eye Tissues. Biochem. J., 51, 521 (1952).

74. DOREA, J.G., and S.E. PEREIRA: The Influence of Hair Color on the Concentration of Zinc and Copper in Boys' Hair. J. Nutrition, 113, 2375 (1983).

75. MOLOKHIA, M., and B. PORTNOY: Trace Elements and Skin Pigmentation. Br. J. Dermatol., 89, 207 (1973).

76. SHIBATA, T., G. PROTA, and Y. MISHIMA: Non-Melanosomal Regulatory Factors in Melanogenesis. J. Invest. Dermatol., 274S (1993).

77. BU'LOCK, J.D., and J. HARLEY-MASON: Melanin and Its Precursors, Part III: New Synthesis of 5,6-Dihydroxyindole and Its Derivatives. J. Chem. Soc., 2248 (1951).

78. NAPOLITANO, A., F. CHIOCCARA, and G. PROTA: A Re-Examination of the Zinc-Catalysed Rearrangement of Dopachrome Using Immobilised Tyrosinase. Gazz. Chim. Ital., 115, 357 (1985).

79. PALUMBO, A., M. D'ISCHIA, G. MISURACA, and G. PROTA: Effect of Metal Ions on the Rearrangement of Dopachrome. Biochim. Biophys. Acta, 925, 203 (1987).

80. KORNER, A., and J. PAWELEK: Dopachrome Conversion: A Possible Control Point in Melanin Biosynthesis. J. Invest. Dermatol., 75, 192 (1980).

81. SUGUMARAN, M.: Letter to the Editor. Pigment Cell Res., 5, 203 (1992).

82. BARBER, J.I., D. TOWNSEND, D.P. OLDS, and R.A. KING: Dopachrome Oxidoreductase: A New Enzyme in the Pigment Pathway. J. Invest. Dermatol., 83, 145 (1984).

83. LEONARD, L.J., D. TOWNSEND, and R.A. KING: Function of Dopachrome Oxidoreductase and Metal Ions in Dopachrome Conversion in the Eumelanin Pathway. Biochemistry, 27, 6156 (1988).

84. PAWELEK, J.M.: Dopachrome Conversion Factor Functions as an Isomerase. Biochem. Biophys. Res. Commun., 166, 1328 (1990).

85. AROCA, P., J.C. GARCIA-BORRON, F. SOLANO, and J.A. LOZANO: Regulation of Distal Mammalian Melanogenesis, I: Partial Purification and Characterization of a Dopachrome Converting Factor: Dopachrome Tautomerase. Biochim. Biophys. Acta, 1035, 266 (1990).

86. DRYJA, T.P., M. O'NEIL-DRYJA, and D.M. ALBERT: Elemental Analysis of Melanins from Bovine Hair, Iris, Choroid and Retinal Pigment Epithelium. Invest. Ophthalmol. Vis. Sci., 18, 231 (1979).

87. PALUMBO, A., M. D'ISCHIA, G. MISURACA, L. DE MARTINO, and G. PROTA: A New Dopachrome-Rearranging Enzyme from the Ejected Ink of the Cuttlefish Sepia officinalis. Biochem. J., 299, 839–844 (1994).

88. TSUKAMOTO, K., I.J. JACKSON, K. URABE, P. MONTAGUE, and V.J. HEARING: A Second Tyrosinase-Related Protein, TRP-2, is a Melanogenic Enzyme Termed Dopachrome Tautomerase. EMBO J. (in press).

89. MASON, H.S.: The Chemistry of Melanin, III: Mechanism of the Oxidation of Dihydroxyphenylalanine by Tyrosinase. J. Biol. Chem., 172, 83 (1948).

90. BU'LOCK, J.D., and J. HARLEY-MASON: Melanin and Its Precursors, Part II: Model Experiments on the Reactions Between Quinones and Indole, and Consideration of a Possible Structure for the Melanin Polymer. J. Chem. Soc., 703 (1951).

91. CROMARTIE, R.I.T., and J. HARLEY-MASON: Melanin and Its Precursors, Part VII: Synthesis of Methylated 5,6-Dihydroxyindoles, β-(4,5-Dihydroxy-2-methylphenyl)-alanine, and Related Amines. J. Chem. Soc., 3525 (1953).

92. CROMARTIE, R.I.T., and J. HARLEY-MASON: Melanin and Its Precursors, Part VIII: The Oxidation of Methylated 5,6-Dihydroxyindoles. Biochem. J., 66, 713 (1957).
93. BEER, R.J.S., T. BROADHURST, and A. ROBERTSON: The Chemistry of the Melanins, Part V: The Autoxidation of 5,6-Dihydroxyindoles. J. Chem. Soc., 1947 (1954).
94. KIRBY, G.W., and L. OGUNKOYA: Structure of Melanin Derived from (±)-3,4-Dihydroxy-(^{14}C,^{3}H)-phenylalanine by Oxidation with Tyrosinase. J. Chem. Soc., Chem. Comm., 21, 546 (1965).
95. NAPOLITANO, A., M.G. CORRADINI, and G. PROTA: A Reinvestigation of the Structure of Melanochrome. Tetrahedron Lett., 26, 2805 (1985).
96. CORRADINI, M.G., A. NAPOLITANO, and G. PROTA: A Biosynthetic Approach to the Structure of Eumelanins. The Isolation of Oligomers from 5,6-Dihydroxy-1-methylindole. Tetrahedron, 42, 2083 (1986).
97. D'ISCHIA, M., A. NAPOLITANO, K. TSIAKAS, and G. PROTA: New Intermediates in the Oxidative Polymerisation of 5,6-Dihydroxyindole to Melanin Promoted by the Peroxidase/H$_2$O$_2$ System. Tetrahedron, 46, 5789 (1990).
98. D'ISCHIA, M., A. NAPOLITANO, and G. PROTA: Peroxidase as an Alternative to Tyrosinase in the Oxidative Polymerization of 5,6-Dihydroxyindoles to Melanin(s). Biochem. Biophys. Acta, 1073, 423 (1991).
99. LAND, E.J.: Pulse Irradiation Studies of Some Reactive Intermediates of Melanogenesis. Revs. Chem. Interm., 10, 219 (1988).
100. AL-KAZWINI, A.T., P. O'NEAL, G.E. ADAMS, R.B. CUNDALL, G. LANG, and A. JUNINO: Reactions of Indolic Radicals Produced Upon One-Electron Oxidation of 5,6-Dihydroxyindole and Its N(1)-Methylated Analogue. J. Chem. Soc. Perkin Trans. 2, 1941 (1991).
101. AL-KAZWINI, A.T., P. O'NEAL, R.B. CUNDALL, G.E. ADAMS, A. JUNINO, and J. MAIGNANT: Direct Observation of the Reaction of the Quinone Methide from 5,6-Dihydroxyindole with the Nucleophilic Azide Ion. Tetrahedron Lett., 33, 3045 (1992).
102. LAMBERT, C., J.N. CHACON, M.R. CHEDEKEL, E.J. LAND, P.A. RILEY, A. THOMPSON, and G. TRUSCOTT: A Pulse Radiolysis Investigation of the Oxidation of Indolic Melanin Precursors: Evidence for Indolequinones and Subsequent Intermediates. Biochim. Biophys. Acta, 993, 12 (1989).
103. ITO, S., and J.A.C. NICOL: Isolation of Oligomers of 5,6-Dihydroxyindole-2-carboxylic Acid from the Eye of the Catfish. Biochem. J., 143, 207 (1974).
104. PALUMBO, P., M. D'ISCHIA, and G. PROTA: Tyrosinase Promoted Oxidation of 5,6-Dihydroxyindole-2-carboxylic Acid to Melanin. Isolation and Characterization of Oligomer Intermediates. Tetrahedron, 43, 4203 (1987).
105. NAPOLITANO, A., O. CRESCENZI and G. PROTA: Copolymerisation of 5,6-Dihydroxyindole and 5,6-Dihydroxyindole-2-carboxylic Acid in Melanogenesis. Isolation of a Cross-Coupling Product. Tetrahedron Lett., 34, 885 (1993).
106. MISURACA, G., G. PROTA, J.T. BAGNARA, and S.K. FROST: Identification of the Leaf-Frog Melanophore Pigment, Rhodomelanochrome, as Pterorhodanin. Comp. Biochem. Physiol., 57B, 41 (1977).
107. FOX, D.L., and D.M. UPPDEGRAFF: Adenochrome, A Glandular Pigment in Branchial Hearts of the Octopus. Arch. Biochem., 1, 339 (1943).
108. PROTA, G.: Nitrogenous Pigments in Marine Invertebrates. In: Marine Natural Products, Vol. 3 (J.P. SCHEUER, ed.), p. 141. New York: Academic Press. 1980.
109. PROTA, G., G. SCHERILLO, O. PETRILLO, and R.A. NICOLAUS: Struttura e biogenesi delle feomelanine, Nota X: Sulla struttura delle tricosiderine. Gazz. Chim. Ital., 93, 1193 (1969).
110. AGRUP, G., C. HANSSON, H. ROBSMAN, A.-M. ROSENGREN, and E. ROSENGREN: 5-S-

Cysteinuldopa and Trichochromes in Red Feathers. Acta Derm. Venereol. (Stockh.), **58**, 269 (1978).

111. RORSMAN, H., P. AGRUP, B. CARLEN, C. HANSSON, N. JONSSON, E. ROSENGREN, and E. TEGNER: Trichochromuria in Melanosis of Melanoma. Acta Derm. Venereol. (Stockh.), **66**, 468 (1986).

112. THOMSON, R.H.: The Pigments of Reddish Hair and Feathers. Angew. Chem. Int. Ed. Engl., **13**, 305 (1974).

113. BROWN, C., and R.M. DAVIDSON: 1,4-Benzothiazines, Dihydro-1,4-benzothiazines and Related Compounds. In: Advances in Heterocyclic Chemistry, Vol. 38 (A.R. KATRITSKY, ed.), p. 135. New York: Academic Press. 1985.

114. NICOLAUS, R.A., G. PROTA, C. SANTACROCE, G. SCHERILLO, and D. SICA: Struttura e biogenesi delle feomelanine, Nota VII: Sulla struttura delle tricosiderine. Gazz. Chim. Ital., **99**, 323 (1969).

115. PROTA, G., A. SUARATO, and R.A. NICOLAUS: The Isolation and Structure of Trichosiderin B. Experientia, **27**, 1381 (1971).

116. KAUL, B.L.: Studies on Heterocyclic Colouring Matters, Part II: $\Delta^{2,2'}$-Bi(2H-1,4-benzothiazines). Helv. Chim. Acta, **57**, 2664 (1974).

117. PROTA, G., E. PONSIGLIONE, and R. RUGGIERO: Synthesis and Photochromism of $\Delta^{2,2'}$-Bi-(2H-1,4-benzothiazine). Tetrahedron, **30**, 2781 (1974).

118. MINALE, L., E. FATTORUSSO, G. CIMINO, S. DE STEFANO, and R.A. NICOLAUS: Struttura e biogenesi delle feomelanine, Nota III: Prodotti di degradazione. Gazz. Chim. Ital., **97**, 1636 (1967).

119. DEIBEL, R.B., and M.R. CHEDEKEL: Biosynthetic and Structural Studies on Pheomelanin. J. Am. Chem. Soc., **104**, 7306 (1982).

120. FATTORUSSO, E., L. MINALE, S. DE STEFANO, G. CIMINO, and R.A. NICOLAUS: Struttura e biogenesi delle feomelanine, Nota V: Sulla Struttura della gallofeomelanina-1. Gazz. Chim. Ital., **98**, 1443 (1968).

121. DEIBEL, R.B.: Biosynthetic and Structural studies on Pheomelanin. Ph.D. Thesis. Baltimore, Maryland: The John Hopkins University. 1983.

122. SEALY, R.C., J.S. HYDE, C.C. FELIX, I.A. MENON, and G. PROTA: Eumelanins and Pheomelanins: Characterization by Electron Spin Resonance Spectroscopy. Science, **217**, 545 (1982).

123. SEALY, R.C., J.S. HIDE, C.C. FELIX, I.A. MENON, G. PROTA, H.M. SWARTZ, S. PERSAD, and H.F. HABERMAN: Novel Free Radical in Synthetic and Natural Pheomelanins: Distinction Between Dopa Melanins and Cysteinyldopa Melanins By ESR Spectroscopy. Proc. Natl. Acad. Sci. (USA), **79**, 2885 (1982).

124. PATIL, D.G., and M.R. CHEDEKEL: Synthesis and Analysis of Pheomelanin Degradation Products. J. Org. Chem., **49**, 997 (1984).

125. FATTORUSSO, E., L. MINALE, G. CIMINO, S. DE STEFANO, and R.A. NICOLAUS: Struttura e biogenesi delle feomelanine, Nota VI: Sulla struttura della gallofeomelanina. Gazz. Chim. Ital., **99**, 29 (1969).

126. FATTORUSSO, E., L. MINALE, G. CIMINO, S. DE STEFANO, and R.A. NICOLAUS: Struttura e biogenesi delle feomelanine, Nota XIII: Sulla struttura della galloeomelanina. Gazz. Chim. Ital., **100**, 880 (1970).

127. MINALE, L., E. FATTORUSSO, G. CIMINO, S. DE STEFANO, and R.A. NICOLAUS: Struttura e biogenesi delle feomelanine, Nota VIII: Sulla struttura della gallofeomelanina-1. Gazz. Chim. Ital., **99**, 431 (1969).

128. MINALE, L., E. FATTORUSSO, S. DE STEFANO, and R.A. NICOLAUS: Struttura e biogenesi delle feomelanine, Nota XI: Uteriori ricerche sulla biogenesi delle feomelanine. Gazz. Chim. Ital., **100**, 461 (1970).

129. FATTORUSSO, E., L. MINALE, and G. SODANO: Feomelanine e eumelanine da nuove fonti naturali. Gazz. Chim. Ital., **100**, 452 (1970).

130. ITO, S., and K. FUJITA: Microanalysis of Eumelanin and Pheomelanin in Hair and Melanosomas by Chemical Degradation and Liquid Chromatography. Anal. Biochem., **144**, 527 (1985).

131. DEIBEL, R.B., and M.R. CHEDEKEL: Biosynthetic and Structural Studies on Pheomelanin, 2. J. Am. Chem. Soc., **106**, 5884 (1984).

132. CHEDEKEL, M.R., K.V. SUBBARAO, P. BAHAN, and T.M. SCHULTZ: Biosynthetic and Structural Studies on Pheomelanin. Biochim. Biophys. Acta, **912**, 239 (1987).

133. PROTA, G., G. SCHERILLO, E. NAPOLANO, and R.A. NICOLAUS: Struttura e biogenesi delle feomelanine, Nota II: Sulla reazione tra o-chinoni e cisteina. Gazz. Chim. Ital., **97**, 1451 (1967).

134. PROTA, G., G. SCHERILLO, and R.A. NICOLAUS: Struttura e biogenesi delle feomelanine, Nota IV: Sintesi e proprieta' della 5-S-cisteinildopa. Gazz. Chim. Ital., **98**, 495 (1968).

135. FATTORUSSO, E., L. MINALE, S. DE STEFANO, G. CIMINO, and R.A. NICOLAUS: Struttura e biogenesi delle feomelanine, Nota IX: Feomelanine biosintetiche. Gazz. Chim. Ital., **99**, 969 (1969).

136. ITO, S., and G. PROTA: A Facile One-Step Synthesis of Cysteinyldopas Using Mushroom Tyrosinase. Experientia, **33**, 1118 (1977).

137. BJORKLUND, A., B. FALCK, S. JACOBSSON, H. RORSMAN, A.-M. ROSENGREN, and E. ROSENGREN: Cysteinyldopa in Human Malignant Melanoma. Acta Derm. Venereol. (Stockh), **52**, 357 (1972).

138. RORSMAN, H., G. AGRUP, C. HANSSON, A.-M. ROSENGREN, and E. ROSENGREN: Detection of Phaeomelanins. In: Pigment Cell: Biologic Basis of Pigmentation, Vol. 4 (S. KLAUS, ed.), p. 244. Basel: Karger. 1979.

139. PROTA, G., H. RORSMAN, A.-M. ROSENGREN, and E. ROSENGREN: Isolation of 2-S-Cysteinyldopa and 2,5,-S,S-Dicysteinyldopa from the Urine of Patients with Melanoma. Experientia, **33**, 720 (1977).

140. RORSMAN, H., G. AGRUP, C. HANSSON, and E. ROSENGREN: Biochemical Recorders of Malignant Melanoma. In: Pigment Cell, Vol. 6 (R.M. MACKIE, ed.), p. 93. Basel: Karger. 1983.

141. ITO, S., and J.A.C. NICOL: A New Amino Acid, 3-(2,5-S,S-Dicysteinyl-3,4-Dihydroxyphenyl)alanine, from the *Tapetum Lucidum* of the Gar (*Lepisosteidae*) and Its Enzymic Synthesis. Biochem. J., **161**, 499 (1977).

142. MORISHIMA, T.F., F. TATSUMI, E. FUKADA, M. SAITO, M. FUJITA, N. NAGASHIMA, and S. HANAWA: Cysteinyldopa Isomers and Dopa in Lesions and Urine of Japanese Patients with Malignant Melanoma. Arch. Dermatol. Res., **275**, 76 (1983).

143. MEISTER, A., and M.E. ANDERSON: Glutathione. Ann. Rev. Biochem., **52**, 711 (1983).

144. PROTA, G., M. D'ISCHIA, and A. NAPOLITANO: The Regulatory Role of Sulfhydryl Compounds in Melanogenesis. Pigment Cell Res., **1S**, 48 (1988).

145. ITO, S., A. PALUMBO and G. PROTA: Tyrosinase-Catalysed Conjugation of Dopa with Glutathione. Experientia, **41**, 960 (1985).

146. RORSMAN, H., E. ALBERTSSON, L.E. EDHOLM, C. HANSSON, L. OGREN, and E. ROSENGREN: Thiols in the Melanocyte. Pigment Cell Res., **1S**, 54 (1988).

147. KARG, E., G. ODH, E. ROSENGREN, A. WITTBJER, and H. RORSMAN: Melanin-Related Biochemistry of IGR 1 Human Melanoma Cells. Melanoma Res., **1**, 5 (1991).

148. FEHLING, C., C. HANSSON, J. POULSEN, H. RORSMAN, and E. ROSENGREN: Formation of Glutathionedopa in Albino Rats After DOPA Injection. Acta Derm. Venereol. (Stockh.), **61**, 339 (1981).

149. ITO, S., and K. FUJITA: New Possible Routes for the Biosynthesis of Cysteinyldopas

and Related Metabolites. In: Pigment Cell 1981: Phenotypic Espression in Pigment Cells (M. SEIJI, ed.), p. 85. Tokyo: University of Tokyo Press. 1981.

150. NKPA, N.N., and M. CHEDEKEL: Mechanistic Studies on the Addition of Cysteine to 3,4-Dihydroxyphenylalanine. J. Org. Chem., **46**, 213 (1981).

151. PALUMBO, A., G. NARDI, M. D'ISCHIA, G. MISURACA, and G. PROTA: Non-Enzymic Oxidation of Cysteinyldopa Catalyzed by Metallic Ions. Gen. Pharmacol., **14**, 253 (1983).

152. CRESCENZI, S., G. MISURACA, E. NOVELLINO, and G. PROTA: Reazioni modello per la biosintesi dei pigmenti feomelanici. Chimica e Industria, **57**, 392 (1975).

153. COSTANTINI, C., O. CRESCENZI, G. PROTA, and A. PALUMBO: New Intermediates of Phaeomelanogenesis in vitro Beyond the 1,4-Benzothiazine Stage. Tetrahedron, **46**, 6831 (1990).

154. YOUNG, T.E., J.R. ORISWOLD, and M.H. HULBERT: Melanin, 1: Kinetics of Oxidative Cyclization of Dopa to Dopachrome. J. Org. Chem., **39**, 1980 (1974).

155. KALYANARAMAN, B., C.C. FELIX, and R.C. SEALY: Semiquinone Anion Radicals of Catecholamines, Catechol Estrogens, and Their Metal Ion Complexes. Environ. Health Perspect., **64**, 185 (1985).

156. CABANES, J., F. GARCIA-CANOVAS, J.A. LOZANO, and F. GARCIA-CARMONA: A Kinetic Study of the Melanization Pathway Between L-Tyrosine and Dopachrome. Biochim. Biophys. Acta, **923**, 187 (1987).

157. RODRIGUEZ-LOPEZ, J.N., J. TUDELA, R. VARON, and F. GARCIA-CANOVAS: Kinetic Study on the Effect of pH on the Melanin Biosynthesis Pathway. Biochim. Biophys. Acta, **1076**, 379 (1991).

158. CLEFFMAN, G.: Function-Specific Changes in the Metabolism of Agouti Pigment Cells. Exp. Cell Res., **35**, 590 (1964).

(*Received May 24, 1993*)

Structure, Occurrence, Biosynthesis, Biological Activity, Synthesis, and Chemistry of Aromadendrane Sesquiterpenoids

H. J. M. Gijsen, J. B. P. A. Wijnberg, and Ae. de Groot, Laboratory of Organic Chemistry, Wageningen Agricultural University, Wageningen, The Netherlands

Contents

1. Structure and Occurrence

Aromadendranes (**1**) belong to a class of sesquiterpenes, structurally characterized by a dimethyl cyclopropane ring fused to a hydroazulene skeleton, as depicted in Fig. 1. Throughout this and the following sections the numbering of the carbon skeleton will be used as given in (**1**). Next to the aromadendranes, the related sesquiterpenes with a 1,7-cycloaromadendrane, 7,8-seco-, and 9,10-secoaromadendrane skeleton will be discussed. Some dimeric and alkylated aromadendranes, *e.g.* macrocarpals and prenylaromadendranes are also included in this review. The literature is covered through September 1993.

The name "aromadendrane" originates from the sesquiterpene (+)-aromadendrene (**2**), the first reported sesquiterpene with this skeleton (*1*).

(1)

(2)

Fig. 1. The aromadendrane skeleton and numbering

(+)-Aromadendrene is a constituent of the essential oil extracted from the wood of *Eucalyptus* trees. In earlier days the trees belonging to this genus were known as *Aromadendron* trees.

After numerous experiments, the structure of (+)-aromadendrene (2) was elucidated in 1953 (2), but it took thirteen years before BÜCHI *et al.* in 1966 established the absolute configuration of 2 through the synthesis of (−)-aromadendrene starting from (−)-perillaldehyde (3) (see Sect. 4). In addition, the absolute configurations of alloaromadendrene (3), in which the hydroazulene skeleton is *cis*-fused, and the related tertiary C7-alcohols globulol (4), epiglobulol (5), ledol (6), and viridiflorol (7) were established.

Gurjun balsam, the resin from tropical trees belonging to the *Dipterocarpaceae* family, provided the aromadendrane (−)-α-gurjunene (25) (4), whose structure was determined in 1963 by OURISSON *et al.* (5). This hydrocarbon can be oxidized to the α,β-unsaturated ketone (−)-cyclocolorenone (27) (6), which has been isolated from *Pseudowintera colorata* (7). The stereochemistry of (27) was independently determined through synthesis of its C8-epimer in 1966 (8). All the compounds mentioned so far are present in numerous plant species. Other frequently occurring aromadendranes are the tertiary C8 and C11 alcohols palustrol (8) and spathulenol (31), respectively, and the hydrocarbon ledene (43), also known as viridiflorene (9).

A number of other aromadendrane hydrocarbons have been isolated from the oil or resin obtained from different tree species. Gurjun balsam contains besides α-gurjunene (25) also its C7 epimer (26) (10). Tolu balsam, obtained from *Myroxylon balsamum* var. *balsamum*, was found to contain the five hydrocarbons (52), (53), (54), (55), and (56) (11). The aromadendrane hydrocarbon β-spathulene (51) is a minor component in the oil of *Schinus Molle* (12).

In many aromadendranes C7 and/or C11 are oxidized. Although less frequently, oxidation at all the other carbon atoms, except C1 and C2, is also observed. Oxygenated aromadendranes are widespread in *Compositae* (13–19), with, amongst others, members of the tribes *Inulae* (20–24), and *Calendulaceae* (25–27). From the latter and from *Pittosporum tobira* (28) glycosylated aromadendranes have been extracted. Other plants that provided oxygenated aromadendranes were *Phebalium squamulosum* (29), *Ferulago antiochia* (30), *Humulus lupulus* (31), and members of the family of *Labiatae* (32, 33).

Ent-aromadendrane sesquiterpenes which have the mirror image carbon skeleton, have been found in the red alga *Laurencia subopposita* (34), soft corals (*Coelenterata, Octocorallia*), marine sponges (*Porifera*) and liverworts (*Hepaticae*) (Table 2). All of the abundant aromadendranes mentioned above have also been found in their enantiomeric forms, but also *ent*-aromadendranes without known antipodes in higher plants have been isolated.

The soft corals *Sinularia mayi* (*35*) and *Cespitularia* sp. aff. *subviridis* (*36,37*) have been reported to contain aromadendranes with both chiralities. In the case of *C. subviridis* the isolated (+)-ledol (**6**) might actually be *ent*-(+)-ledol (**61**), because the sign of rotation is dependent on the solvent used (*38*). *Ent*-aromadendranes with a new structure have been isolated from the soft corals *Clavularia koellikeri* (*39*) and *Xenia novae-brittaniae* (*40*).

Some marine sponges contain aromadendranes with isonitrile, isothiocyanate, or formamide groups at C7 or C8. Epipolasin B (**78**) has been isolated from *Epipolasis kushimotoensis* (*41*). The same structure has been proposed for axisothiocyanate-2, isolated from *Axinella cannabina* (*42*) and *Acanthella pulcherrima* (*43*). Epipolasin B (**78**) has been proved to possess the *ent*-configuration (*41*). The absolute stereochemistry of the aromadendranes from *Axinella cannabina*, including axisothiocyanate-2 is not known, but the *ent*-configuration is most likely. Aromadendranes with a nitrogen containing group at C8 have been found in *Acanthella* species (*43, 44*). The stereochemistry at C7 and C8, at first unclear, was proved to be β as indicated in (**84**), (**85**) and (**86**) (*37a*). Axisonitrile-2 (**77**) and axisothiocyanate-2 (**78**), found in the nudibranch *Cadlina luteomarginata*, probably originate from the sponge diet of this mollusc (*45*).

All aromadendranes found in liverworts (*Hepaticae*) possess the *ent*-configuration. A number of *ent*-aromadendranes, all of which have known antipodal counterparts in higher plants, have been isolated from liverworts (*46–51*). In a few cases, their antipodes have not been found in higher plants (*24, 50, 52*). From the liverwort genus *Plagiochila* many *ent*-9,10-secoaromadendranes have been isolated (*48, 53–60*) (Table 3). The liverwort *Mylia taylorii* is a rich source of unusual sesquiterpenoids derived from *ent*-aromadendranes. Besides *ent*-aromadendranes, several *ent*-7,8-secoaromadendranes (*47, 61*), an *ent*-nor-7,8-secoaromadendrane (**113**) (*62*), and several *ent*-1,7-cycloaromadendranes (*47, 63, 64*) have been isolated (Tables 4 and 5). Anastreptene (**118**), one of the *ent*-1,7-cycloaromadendranes is a common constituent in many liverwort species (*65*).

Three dimeric sesquiterpenoids, isolated from *Mylia taylorii*, are Diels-Alder-type adducts of the *ent*-7,8-secoaromadendrane (−)-taylorione (**110**) and the *ent*-1,7-cycloaromadendrane (−)-myliol (**114**) (*66*). The dimeric *ent*-spiroterpenoid plagiospirolide E (**122**) has been isolated from *Plagiochila moritziana* (*67*). This C_{30}-terpenoid is composed of an aromadendrane and an eudesmane unit. Various other aromadendranes with extra alkyl substituents have been isolated (Table 6). Tanzanene (123) is a C7-spiro benzopyranyl aromadendrane, isolated from *Uvaria tanzaniae* (68). Prenylaromadendranes (cneorubins), consisting of an aromadendrane unit with an isopentylgroup at C13, have been isolated from *Cneorum tricoccon* (69). *Eucalyptus globulus* (70, 71) and *E. macrocarpa* (72, 73) contain aromadendranes in which C11 bears an isopentyl phloroglucinol group (macrocarpals or euglobals).

Waitzia acuminata (*Compositae, Inulae*) contains a fragmentated aromadendrane sesquiterpene, waitziacuminone (132), probably derived from spathulenol (31) (22) (Scheme 1).

Scheme 1. The proposed biosynthesis of waitziacuminone

The aromadendranes known so far and the organisms from which they have been isolated for the first time, are listed in Tables 1 to 6. The accompanying structures are given below the respective tables. For the structures of the *ent*-aromadendranes with known antipodal counterparts the reader is referred to the structures of the latter.

Table 1. *Natural Occurring Aromadendranes*

	Compound	Isolated from	Ref.
(2)	(+)-Aromadendrene	*Eucalyptus* species	(1)
(3)	(−)-Alloaromadendrene	*Eucalyptus* species	(74)
(4)	(−)-Globulol	*Eucalyptus globulus*	(75)
(5)	(−)-Epiglobulol	*Humulus lupulus*	(31)
(6)	(−)-Ledol[a]	*Ledum palustre*	(76)
(7)	(+)-Viridiflorol	*Melaleuca viridiflora*	(77)
(8)	(−)-Palustrol	*Ledum palustre*	(76)
(9)	(−)-8-Hydroxy-alloaromadendrene	*Cassinia subtropica*	(21)
(10)	Aromadendrene epoxide	*Humulus lupulus*	(31)
(11)	Alloaromadendrene epoxide	*Humulus lupulus*	(31)
(12)	(−)-Arvoside B	*Calendula arvensis*	(25)
(13)	(−)-Ledol-β-D-fucopyranoside-2′-O-acetate	*Calendula arvensis*	(25)

Table 1 (*continued*)

	Compound	Isolated from	Ref.
(14)	(−)-Ledol-β-D-fucopyranoside-2′-O-2-methylbutyrate	*Calendula arvensis*	*(25)*
(15)	(−)-Ledol-β-D-fucopyranoside-2′-O-4-methylsenecioate	*Calendula arvensis*	*(25)*
(16)	(−)-Ledol-β-D-fucopyranoside-2′-O-isobutyrate	*Calendula arvensis*	*(27)*
(17)	(−)-Ledol-β-D-fucopyranoside-2′-O-angelate	*Calendula arvensis*	*(27)*
(18)	Viridiflorol-β-D-fucopyranoside	*Calendula persica*	*(26)*
(19)	Viridiflorol-β-D-chinovopyranoside	*Calendula persica*	*(26)*
(20)	Viridiflorol-β-D-fucopyranoside-2′-O-4-methylsenecioate	*Calendula persica*	*(26)*
(21)	Viridiflorol-β-D-fucopyranoside-2′-O-senecioate	*Calendula persica*	*(26)*
(22)	Viridiflorol-β-D-chinovopyranoside-2′-O-senecioate	*Calendula persica*	*(26)*
(23)	(+)-Pittosporanoside A1	*Pittosporum tobira*	*(28)*
(24)	(+)-Pittosporanoside A2	*Pittosporum tobira*	*(28)*
(25)	(−)-α-Gurjunene	Gurjun balsam	*(4, 5)*
(26)	7-*epi*-α-Gurjunene	Gurjun balsam	*(10)*
(27)	(−)-Cyclocolorenone	*Pseudowintera colorata*	*(7)*
(28)	5α-Hydroxy-α-gurjunene	*Helichrysum nudifolium*	*(14)*
(29)	(−)-5α-Acetoxy-α-gurjunene	*Helichrysum nudifolium*	*(14)*
(30)	(−)-5β-Acetoxy-α-gurjunene	*Helichrysum nudifolium*	*(14)*
(31)	(+)-Spathulenol	*Eucalyptus spathulata*	*(78)*
(32)	(+)-5α-Hydroxy-spathulenol	*Cineraria fruticulorum,* *Parthenium argentatum*	*(24)* *(16)*
(33)	(−)-Guayulin C	*Parthenium argentatum*	*(18)*
(34)	Guayulin D	*Parthenium argentatum*	*(18)*
(35)	5α-Benzoyloxy-spathulenol	*Ferulago antiochia*	*(30)*
(36)	(−)-6β-Hydroxy-spathulenol	*Sideritis varoi*	*(33)*
(37)	(+)-6β-Acetoxy-spathulenol	*Sideritis varoi*	*(33)*
(38)	(−)-5α,14-Dihydroxy-spathulenol	*Cineraria fruticulorum*	*(24)*
(39)	(−)-5α-Hydroxy-14-oxo-spathulenol	*Cineraria fruticulorum*	*(24)*
(40)	(−)-Aromadendrane-7α,11β-diol	*Brasilia sickii*	*(15)*
(41)	(+)-Alloaromadendrane-7β,11β-diol[b]	*Ambrosia peruviana* *Ambrosia elatior*	*(19a)* *(19b)*
(42)	(−)-Aromadendrane-7α,11α-diol	*Sinularia mayi*	*(35)*
(43)	(+)-Ledene (viridiflorene)	*Melaleuca alternifolia*	*(9)*
(44)	(+)-Isospathulenol	*Salvia sclarea*	*(32)*
(45)	(−)-Squamulosone	*Phebalium squamulosum*	*(29)*
(46)	13,14-Diacetoxy-9-oxo-ledene	*Gnephosis brevifolia*	*(20)*
(47)	(+)-15-Hydroxy-viridiflorol	*Wyethia arizonica,* *Pulicaria paludosa*	*(17)* *(23)*
(48)	(+)-14,15-Dihydroxy-virdiflorol	*Wyethia arizonica*	*(17)*

Table 1 (*continued*)

	Compound	Isolated from	Ref.
(49)	Flourensadiol	*Flourensia cernua*	(*13*)
(50)	(+)-10-Oxo-viridiflorol	*Helichrysum albirosulatum*	(*14*)
(51)	β-Spathulene	*Schinus molle*	(*12*)
(52)	(−)-8-*epi*-α-Gurjunene	*Myroxylon balsamum*	(*11*)
(53)	(+)-8(9)-Aromadendrene	*Myroxylon balsamum*	(*11*)
(54)	(−)-6(7)-Alloaromadendrene	*Myroxylon balsamum*	(*11*)
(55)	1(8)-10(11)-Aromadendradiene	*Myroxylon balsamum*	(*11*)
(56)	1(11)-7(8)-Aromadendranediene	*Myroxylon balsamum*	(*11*)

[a] The sign of rotation of ledol is dependent on the solvent used (*38*).
[b] Revised structure, see (*79*) and (*97*).

(2): $R^1 = \alpha H$ $R^2 = R^3 = CH_2$
(3): $R^1 = \beta H$ $R^2 = R^3 = CH_2$
(4): $R^1 = \alpha H$ $R^2 = CH_3$ $R^3 = OH$
(5): $R^1 = \alpha H$ $R^2 = OH$ $R^3 = CH_3$
(6): $R^1 = \beta H$ $R^2 = CH_3$ $R^3 = OH$
(7): $R^1 = \beta H$ $R^2 = OH$ $R^3 = CH_3$
(8): $R^1 = \alpha OH$ $R^2 = H$ $R^3 = CH_3$
(9): $R^1 = \beta OH$ $R^2 = R^3 = CH_2$

(10): 8αH
(11): 8βH

7α-glycosyl,7β-methyl:
(12): $R^1 = H$ $R^2 = H$ $R^3 = \beta OH$
(13): $R^1 = Ac$ $R^2 = H$ $R^3 = \beta OH$
(14): $R^1 = 2\text{-MeBu}$ $R^2 = H$ $R^3 = \beta OH$
(15): $R^1 = MeSen$ $R^2 = H$ $R^3 = \beta OH$
(16): $R^1 = Bu$ $R^2 = H$ $R^3 = \beta OH$
(17): $R^1 = Ang$ $R^2 = H$ $R^3 = \beta OH$

7α-methyl,7β-glycosyl:
(18): $R^1 = H$ $R^2 = H$ $R^3 = \beta OH$
(19): $R^1 = H$ $R^2 = H$ $R^3 = \alpha OH$
(20): $R^1 = MeSen$ $R^2 = H$ $R^3 = \beta OH$
(21): $R^1 = Sen$ $R^2 = H$ $R^3 = \beta OH$
(22): $R^1 = Sen$ $R^2 = H$ $R^3 = \alpha OH$
(23): $R^1 = Ac$ $R^2 = Ang$ $R^3 = \beta OH$
(24): $R^1 = Ac$ $R^2 = H$ $R^3 = \beta Ang$

(25): $7\alpha CH_3$ $R^1 = H_2$ $R^2 = H$
(26): $7\beta CH_3$ $R^1 = H_2$ $R^2 = H$
(27): $7\alpha CH_3$ $R^1 = O$ $R^2 = H$
(28): $7\alpha CH_3$ $R^1 = H_2$ $R^2 = \alpha OH$
(29): $7\alpha CH_3$ $R^1 = H_2$ $R^2 = \alpha OAc$
(30): $7\alpha CH_3$ $R^1 = H_2$ $R^2 = \beta OAc$

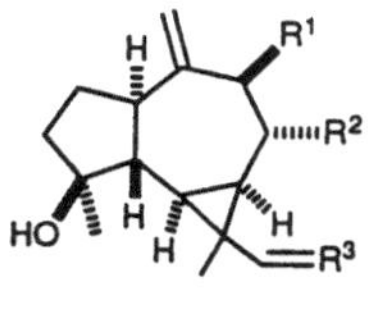

(31): $R^1 = H$ $R^2 = H$ $R^3 = H_2$
(32): $R^1 = H$ $R^2 = OH$ $R^3 = H_2$
(33): $R^1 = H$ $R^2 = OCinn$ $R^3 = H_2$
(34): $R^1 = H$ $R^2 = OAnis$ $R^3 = H_2$
(35): $R^1 = H$ $R^2 = OBz$ $R^3 = H_2$
(36): $R^1 = OH$ $R^2 = H$ $R^3 = H_2$
(37): $R^1 = OAc$ $R^2 = H$ $R^3 = H_2$
(38): $R^1 = H$ $R^2 = OH$ $R^3 = OH,H$
(39): $R^1 = H$ $R^2 = OH$ $R^3 = O$

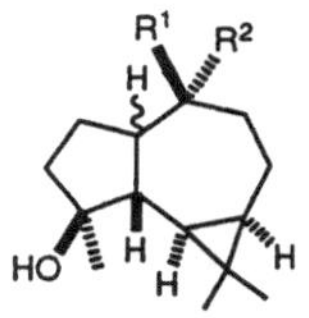

(40): αH $R^1 = CH_3$ $R^2 = OH$
(41): βH $R^1 = OH$ $R^2 = CH_3$

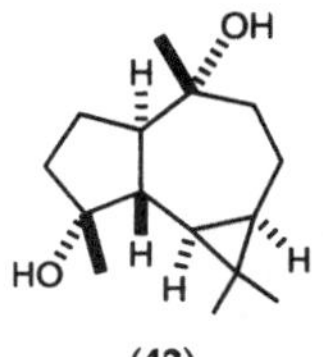

(42)

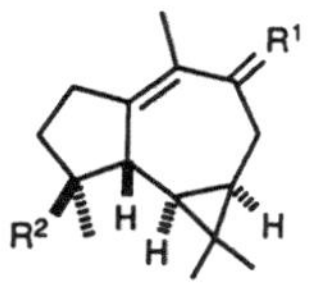

(43): $R^1 = H_2$ $R^2 = H$
(44): $R^1 = H_2$ $R^2 = OH$
(45): $R^1 = O$ $R^2 = H$

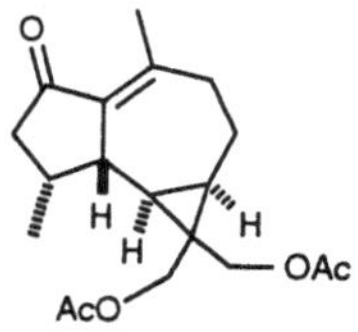

(46)

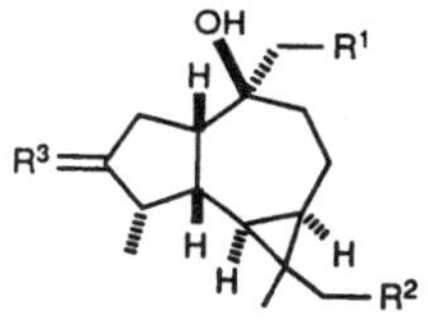

(47): $R^1 = OH$ $R^2 = H$ $R^3 = H_2$
(48): $R^1 = OH$ $R^2 = OH$ $R^3 = H_2$
(49): $R^1 = H$ $R^2 = OH$ $R^3 = H_2$
(50): $R^1 = H$ $R^2 = H$ $R^3 = O$

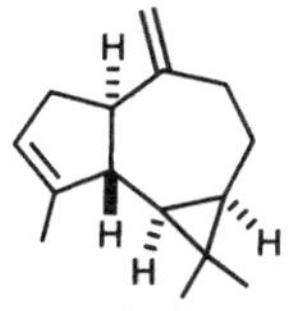

(51)

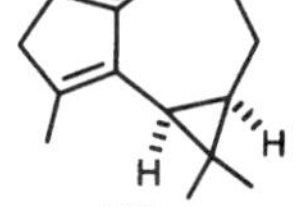

(52)

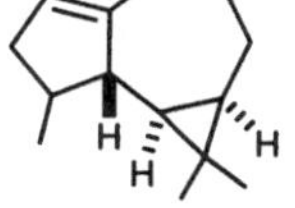

(53)

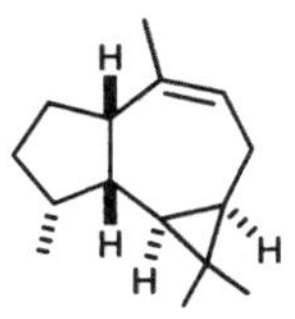

(54)

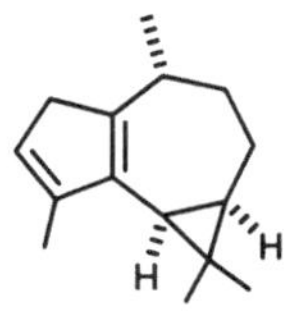

(55)

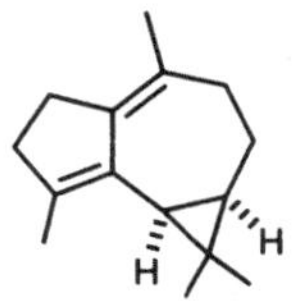

(56)

Table 2. *ent-Aromadendranes*

Compound	Isolated from	Ref.
(57)[a] (−)-Aromadendrene	*Scapania ornithopodioides*	*(46)*
(58)[a] Alloaromadendrene	*Cespitularia* sp. aff. *subviridis*	*(36)*
(59)[a] (+)-Globulol	*Mylia taylorii*	*(47)*
(60)[a] (+)-Epiglobulol	*Plagiochila yokogurensis*	*(48)*
	Diplophyllum albicans	*(49)*
(61)[a] (+)-Ledol	*Cespitularia* sp. aff. *subviridis*	*(36)*
(62)[a] (−)-Viridiflorol	*Cespitularia* sp. aff. *subviridis*	*(36)*
(63)[a] (+)-Palustrol	*Cespitularia* sp. aff. *subviridis*	*(37)*
(64)[a] (−)-Spathulenol	*Plagiochila yokogurensis*	*(48)*
(65)[a] (−)-Ledene	*Plagiochila yokogurensis*	*(48)*
(66)[a] (+)-α-Gurjunene	*Porella* species	*(50)*
(67)[a] (+)-Cyclocolorenone	*Plagiochila acanthophylla*	*(51)*
(68) (+)-8-Hydroxy-cyclocolorenone	*Porella* species	*(50)*
(69) (+)-11(12)-Alloaromadendrene	*Porella* species	*(50)*
(β-gurjunene)[b]		
(70) (+)-11(12)-Dehydroledol	*Mylia taylorii*	*(47)*
(71) (+)-11(12)-Dehydroglobulol	*Mylia taylorii*	*(47)*
(72)[a] (+)-Aromadendrane-7β,11α-diol	*Sinularia mayi*	*(35)*
(73)[a] (−)-Alloaromadendrane-7α,11α-diol[c]	*Sinularia mayi*	*(35)*
(74) (−)-Tridensenone	*Bazzania tridens*	*(52)*
(75)[a] (+)-8-Hydroxy-alloaromadendrene	*Laurencia subopposita*	*(34)*
(76) (−)-8(9)-Dehydroglobulol	*Laurencia subopposita*	*(34)*
(77) (+)-Axisonitrile-2	*Axinella cannabina*	*(42c)*
	Cadlina luteomarginata	*(45)*
(78) (+)-Axisothiocyanate-2	*Axinella cannabina*	*(42b)*
(+)-Epipolasin-B	*Epipolasis kushimotoensis*	*(41)*
	Cadlina luteomarginata	*(45)*
(79) (+)-Axamide-2	*Axinella cannabina*	*(42b)*
(80) Epipolasinthiourea-B	*Epipolasis kushimotoensis*	*(41)*
(81) (−)-7β-Isocyanoalloaromadendrane	*Axinella cannabina*	*(42a)*
(82) (−)-7β-Isothiocyanate-	*Axinella cannabina*,	*(42a)*
alloaromadendrane	*Acanthella pulcherrima*	*(43)*
(83) 7β-Formamidoalloaromadendrane	*Axinella cannabina*	*(42a)*
(84) (−)-8-Isocyano-7β-aromadendrane	*Acanthella acuta*	*(44b,c)*
	A. pulcherrima	*(43)*
(85) (−)-8-Isothiocyanate-	*Acanthella acuta*	*(44b)*
7β-aromadendrane		
(86) 8-Isocyanate-7β-aromadendrane	*Acanthella acuta*	*(44a)*
(87) See structure	*Clavularia koellikeri*	*(39)*
(88) (+)-1(8)-Aromadendren-4-ol	*Xenia novae-brittaniae*	*(40)*

[a] Compound with known antipode, for structure see antipodal structure, Table 1.
[b] The name β-gurjunene for **(69)** is incorrect. This name has already been given to 1(10)-aristolene (calarene).
[c] Revised structure, see *(97)*.

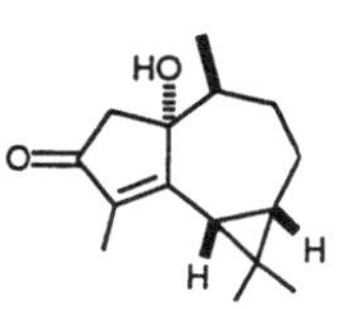

(68)

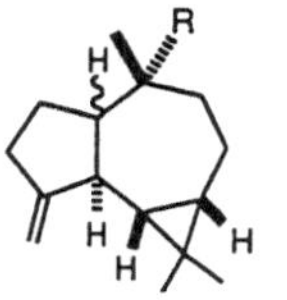

(69): αH R = H
(70): αH R = OH
(71): βH R = OH

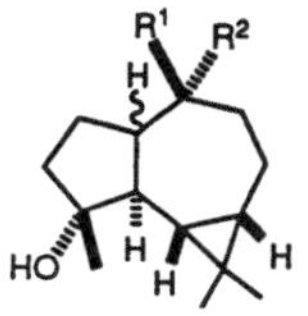

(72): βH R¹ = OH R² = CH₃
(73): αH R¹ = CH₃ R² = OH

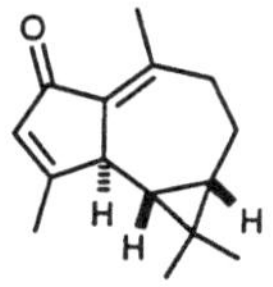

(74)

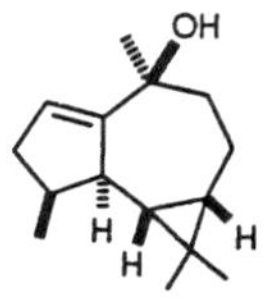

(76)

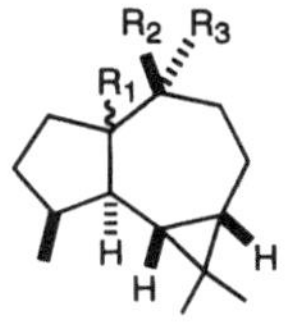

R₂ R₃

(77): R¹ = βH R² = NC R³ = CH₃
(78): R¹ = βH R² = NCS R³ = CH₃
(79): R¹ = βH R² = NHCHO R³ = CH₃
(80): R¹ = βH R² = NHCSNHC₂H₄Ph
       R³ = CH₃
(81): R¹ = αH R² = NC R³ = CH₃
(82): R¹ = αH R² = NCS R³ = CH₃
(83): R¹ = αH R² = NHCHO R³ = CH₃
(84): R¹ = βNC R² = CH₃ R³ = H
(85): R¹ = βNCS R² = CH₃ R³ = H
(86): R¹ = βNCO R² = CH₃ R³ = H

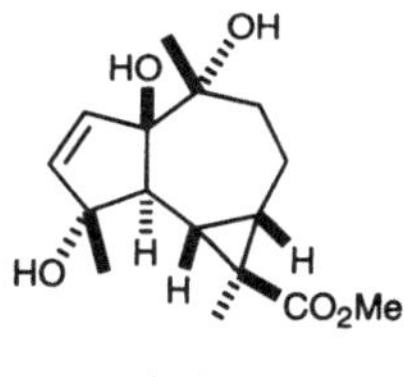

(87)

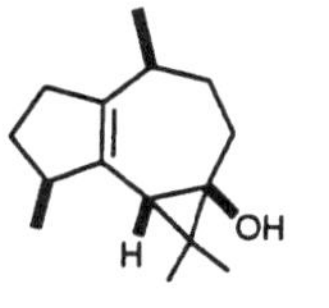

(88)

Table 3. *ent-9,10-Secoaromadendranes*

	Compound	Isolated from	Ref.
(89)	(−)-Plagiochilide	*Plagiochila yokogurensis*	(53)
(90)	(+)-Plagiochiline A	*Plagiochila yokogurensis*	(53)
(91)	(+)-Plagiochiline B	*Plagiochila hattoriana*	(54)
(92)	(+)-Plagiochiline D	*Plagiochila asplenioides*	(57)
(93)	(+)-Plagiochiline G	*Plagiochila ovalifolia*	(48)
(94)	(+)-Plagiochiline I	*Plagiochila yokogurensis*	(48)
(95)	(+)-Plagiochiline E	*Plagiochila asplenioides*	(57)

Table 3 (continued)

	Compound	Isolated from	Ref.
(96)	(+)-Plagiochiline C	*Plagiochila asplenioides*	*(57)*
	(+)-Ovalifoliene	*Plagiochila semidecurrens*	*(55)*
(97)	(+)-Plagiochiline H	*Plagiochila yokogurensis*	*(48)*
	(+)-Deacetoxyovalifoliene	*Plagiochila semidecurrens*	*(59)*
(98)	(+)-6α-Acetoxyovalifoliene	*Plagiochila semidecurrens*	*(59)*
(99)	(+)-Ovalimethoxy I	*Plagiochila semidecurrens*	*(59)*
(100)	(+)-Ovalimethoxy II	*Plagiochila semidecurrens*	*(59)*
	(+)-Methoxyplagiochiline Cᵃ	*Plagiochila yokogurensis*	*(48)*
(101)	(+)-Plagiochiline F	*Plagiochila asplenioides*	*(57)*
(102)	(+)-Ovalifolienal	*Plagiochila semidecurrens*	*(59)*
(103)	(+)-Ovalifolienalone	*Plagiochila semidecurrens*	*(56)*
(104)	(−)-Plagiochiline J	*Plagiochila fruticosa*	*(60)*
(105)	Plagiochiline K	*Plagiochila fruticosa*	*(60)*
(106)	(−)-Furanoplagiochilal	*Plagiochila hattoriana*	*(58)*
(107)	(−)-Plagiochilal A	*Plagiochila hattoriana*	*(58)*
	(−)-Hanegokedial	*Plagiochila semidecurrens*	*(55)*
(108)	(−)-Plagiochilal B	*Plagiochila fruticosa*	*(60)*
(109)	(+)-Hanegoketrial	*Plagiochila semidecurrens*	*(59)*

ᵃ According to Asakawa *et al.* the 10-methoxy compounds are artefacts, formed during isolation *(48)*.

(89)

(90): R¹ = OAc   R² = H   R³ = H
(91): R¹ = OAc   R² = OAc   R³ = H
(92): R¹ = OAc   R² = OAc   R³ = OAc
(93): R¹ = OAc   R² = OAc   R³ = OH
(94): R¹ = OH   R² = H   R³ = H

(95)

(96): R¹ = OAc   R² = H
(97): R¹ = H   R² = H
(98): R¹ = OAc   R² = OAc

(99): βOMe
(100): αOMe

(101)

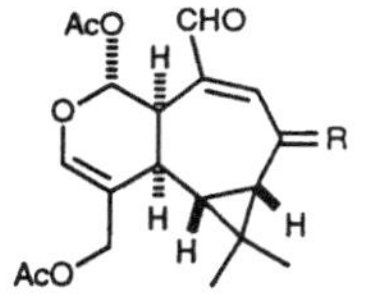

(102): R = H₂
(103): R = O

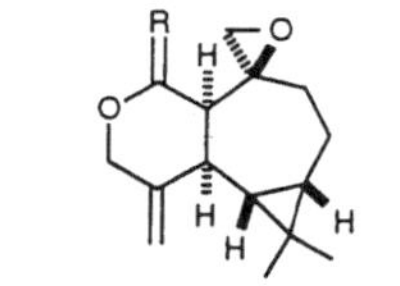

(104): R = O
(105): R = αOH, βH

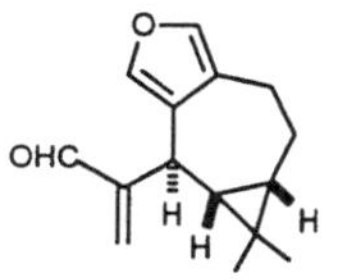

(106)

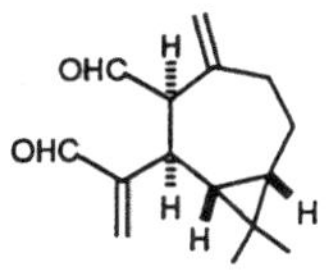

(107)

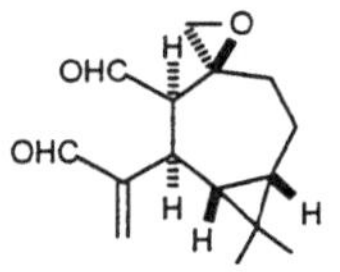

(108)

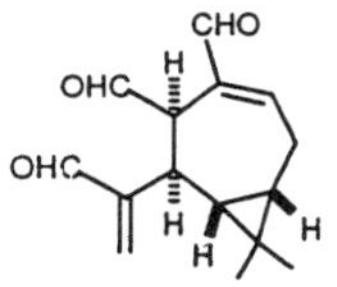

(109)

Table 4. *ent-7,8-Secoaromadendranes*

	Compound	Isolated from	Ref.
(110)	(−)-Taylorione	*Mylia taylorii*	(61)
(111)	(−)-10-Acetoxytaylorione	*Mylia taylorii*	(47)
(112)	(−)-7,10-Dioxotaylori-11-ene	*Mylia taylorii*	(47)
(113)	See structure	*Mylia taylorii*	(62)

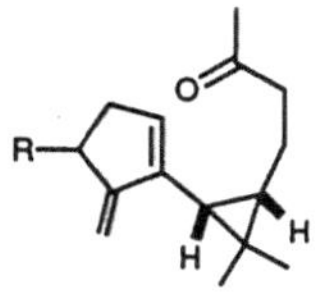

(110): R = H
(111): R = OAc

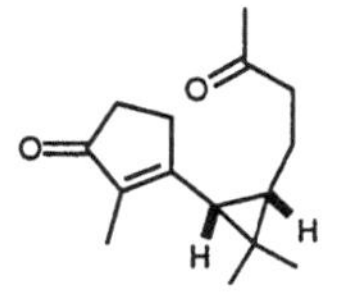

(112)

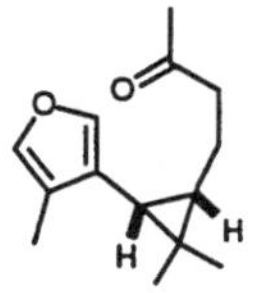

(113)

Table 5. *ent-1,7-Cycloaromadendranes*

	Compound	Isolated from	Ref.
(114)	(−)-Myliol	*Mylia taylorii*	(63)
(115)	(−)-10-Epimyliol	*Mylia taylorii*	(47)
(116)	(+)-Myli-11(12)-en-6-one	*Mylia taylorii*	(47)
(117)	(−)-Dihydromylione A	*Mylia taylorii*	(64)
(118)	(+)-Anastreptene	*Anastrepta orcadensis*	(65)

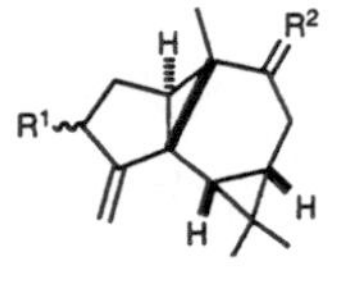

(114): R¹ = βOH   R² = H₂
(115): R¹ = αOH   R² = H₂
(116): R¹ = H   R² = O

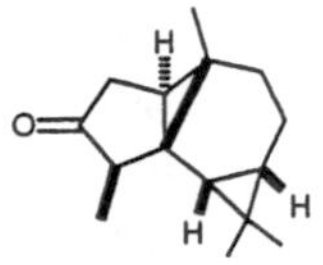

(117)

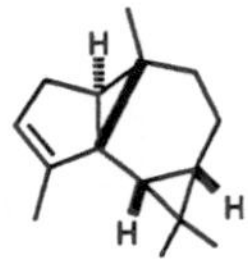

(118)

Table 6. *Dimeric and Alkylated Aromadendranes*

	Compound	Isolated from	Ref.
(119)	Myltaylorione A	*Mylia taylorii*	(66)
(120)	Myltaylorione B	*Mylia taylorii*	(66)
(121)	Bitaylorione	*Mylia taylorii*	(66)
(122)	*ent*-(+)-Plagiospirolide E	*Plagiochila moritziana*	(67)
(123)	(−)-Tanzanene	*Uvaria tanzaniae*	(68)
(124)	(+)-Cneorubine U	*Cneorum tricoccon*	(69)
(125)	Cneorubine W1	*Cneorum tricoccon*	(69)
(126)	Cneorubine W2	*Cneorum tricoccon*	(69)
(127)	(−)-Cneorubine X	*Cneorum tricoccon*	(69)
(128)	(−)-Euglobal V	*Eucalyptus globulus*	(70)
(129)	(−)-Macrocarpal A	*Eucalyptus macrocarpa*	(72)
(130)	(−)-Macrocarpal B	*Eucalyptus globulus*	(71)
		Eucalyptus macrocarpa	(73)
(131)	(−)-Macrocarpal C	*Eucalyptus globulus*	(71)
	(−)-Macrocarpal Gª	*Eucalyptus macrocarpa*	(73)

ª Macrocarpal C from *E. globulus* and macrocarpal G from *E. macrocarpa* are probably identical.

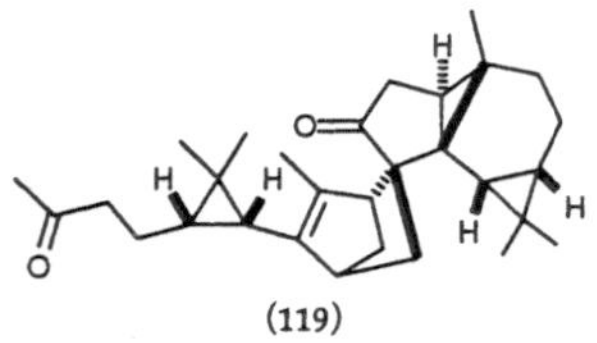

(119)

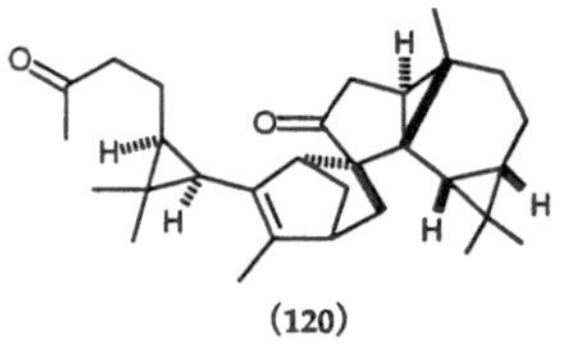

(120)

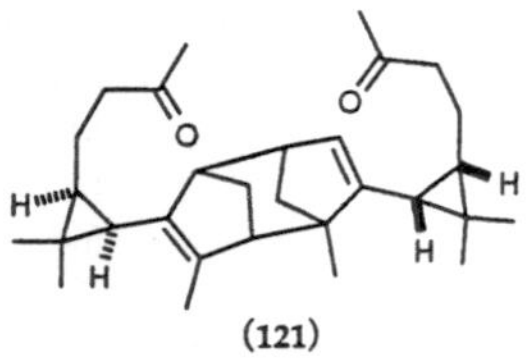

(121)

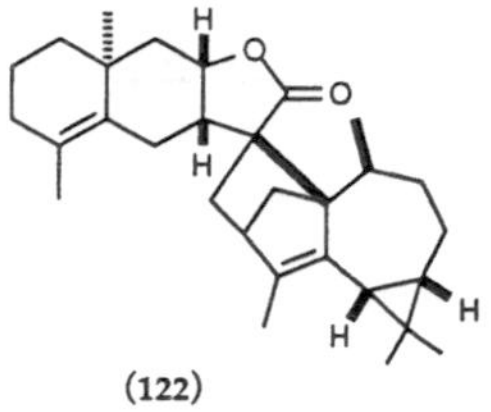

(122)

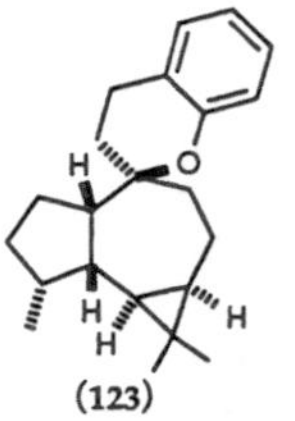

(123)

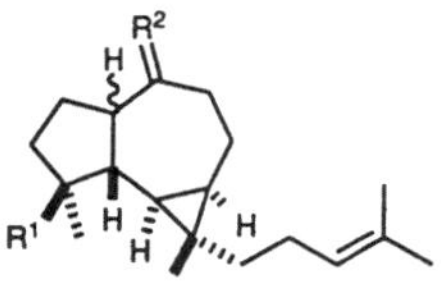

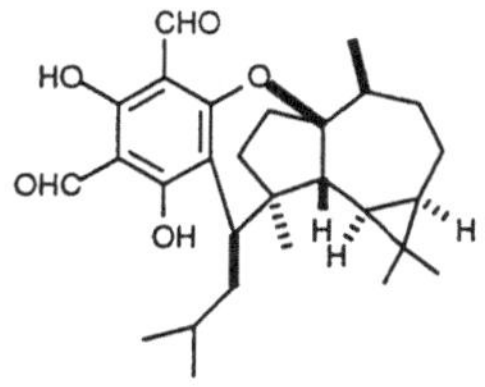

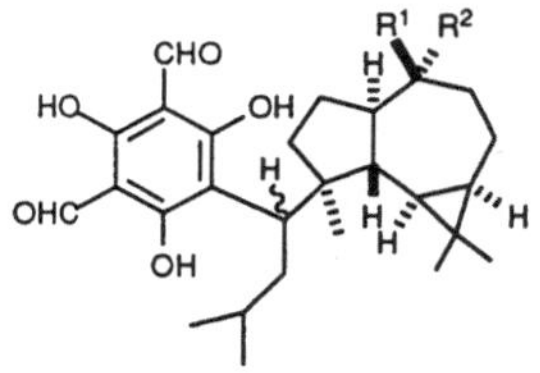

(124): αH R¹ = H   R² = βCH₃,αOH   (128)   (129): αH R¹ = OH   R² = CH₃
(125): βH R¹ = H   R² = CH₂   (130): βH R¹ = OH   R² = CH₃
(126): αH R¹ = H   R² = CH₂   (131): αH R¹ = R² = CH₂
(127): βH R¹ = OH R² = CH₂

2. Biosynthesis

Almost all structural classes of sesquiterpenes can be derived from
trans-farnesylpyrophosphate (*trans*-FPP), and the *cis*-isomer (*cis*-FPP),
through appropriate cyclizations and rearrangements. Enzymatic regiospecific cyclizations of *trans*-FPP and *cis*-FPP generate various mono-
carbocyclic cations through the intermediacy of non-classical carbocations (*80*). The non-classical cation (133) obtained from *trans*-FPP
gives, after a stereoselective 1,3-deprotonation, the bicyclogermacrene
(134) (Scheme 2). The latter compound is believed to be the biosynthetic
precursor of aromadendranes and maalianes (*81–83*). Treatment of (134)
with acid gives, among other products, the aromadendrane sesquiterpene
(+)-ledene (43) (*82*). In many plants containing aromadendranes (+)-
bicyclogermacrene (134) has also been identified (*84*). From organisms

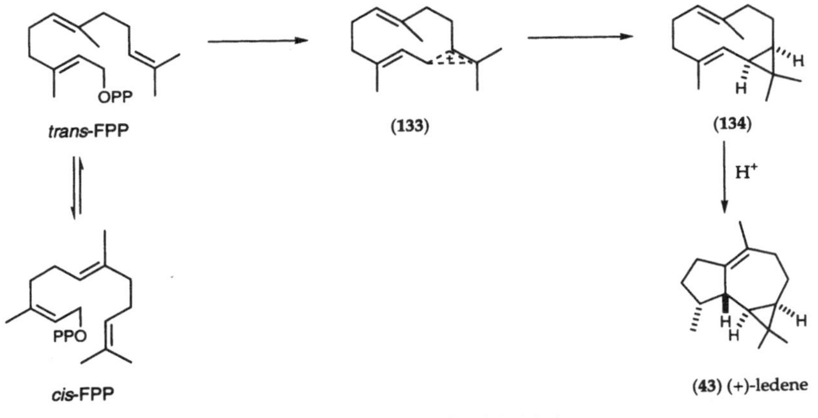

Scheme 2. The biosynthesis of (+)-ledene

that produce *ent*-aromadendranes often (−)-*ent*-bicyclogermacrene has been isolated (*85*). The absolute stereochemistry of an aromadendrane is therefore determined by the enzymes that produce either (+)-bicyclo-germacrene (**134**) or its (−)-enantiomer.

A transannular cyclization of the most stable conformer (**134a**) of bicyclogermacrene (*86*) would produce the *cis*-fused alloaromadendrane carbocations (**135**) or (**137**) after electrophilic attack at C7 or C11, respectively (*82*) (Scheme 3). Similarly the *trans*-fused maaliane car-bocation (**136**) can be formed after electrophilic attack at C8.

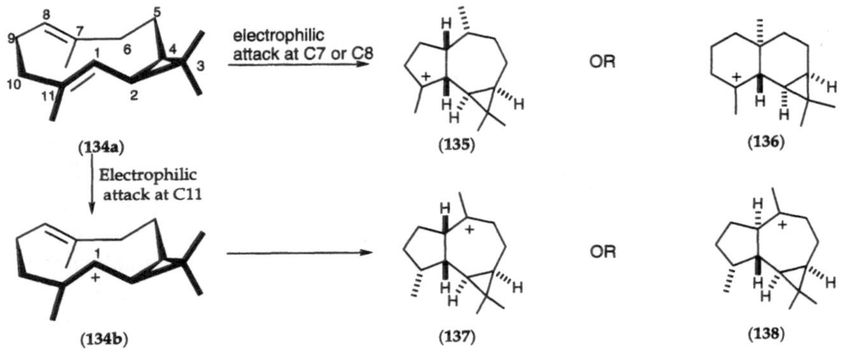

Scheme 3. The biosynthesis of *cis*- and *trans*-fused aromadendranes

The formation of the *trans*-fused aromadendrane carbocation (**138**) from (**134**) is more difficult to explain. A cyclization similar to that described for the alloaromadendranes would require a less stable con-former of (**134**) (*83*). However, attack of an electrophile at C11, or acid-catalyzed opening of a C1–C11 epoxide in bicyclogermacrene (**134**) in anti-Markownikoff fashion would lead to a cyclopropyl carbinyl cation at C1 (**134b**). With this relatively stable carbocation one can imagine cyclization towards *trans*-aromadendranes which are in general thermo-dynamically favoured over their corresponding *cis*-(allo)aromaden-dranes (*87*).

Trans-aromadendranes can be produced from bicyclogermacrenes, as has been proved in the synthesis of guayulin C (**33**) and D (**34**) from guayulin A (**139**) and B (**140**), respectively, via epoxidation with *m*CPBA (*18*) (Scheme 4).

Another theory explains the formation of *trans*-aromadendranes as starting from isobicyclogermacrene (**142**) (*83*). This compound can be

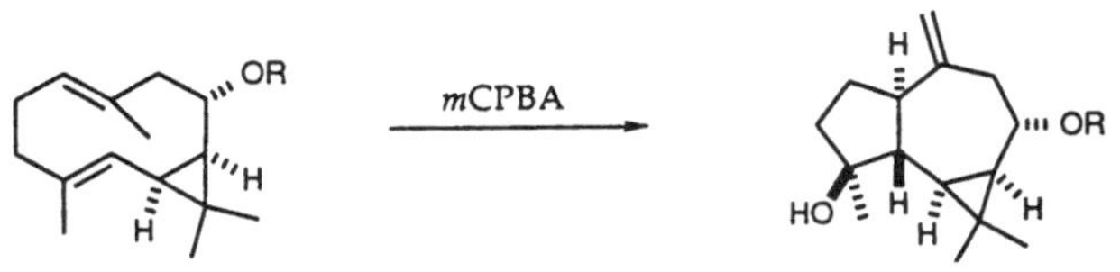

(139): R = Cinnamoyl
(140): R = Anisoyl

(33): R = Cinnamoyl
(34): R = Anisoyl

Scheme 4. The synthesis of *trans*-aromadendranes from bicyclogermacranes

derived from *cis*-FPP via 1,3-deprotonation of the non-classical cation (141), in a way similar to the formation of bicyclogermacrene (134) from *trans*-FPP (Scheme 5). Also it can be formed via isomerization of (134) (*82*). Cyclization of the most stable conformer of (142) should give rise to *trans*-aromadendranes or *cis*-maalianes (*83*). The fact that *cis*-maalianes have never been found in nature speaks against this cyclization process. Furthermore, till now only two isobicyclogermacrenes have been reported to occur in one or two plant species: (+)-isobicyclogermacrenal (143) (*88*) and its (−)-enantiomer (144) (*83*). It is therefore unlikely that all the *trans*-aromadendranes are derived from isobicyclogermacrene (142).

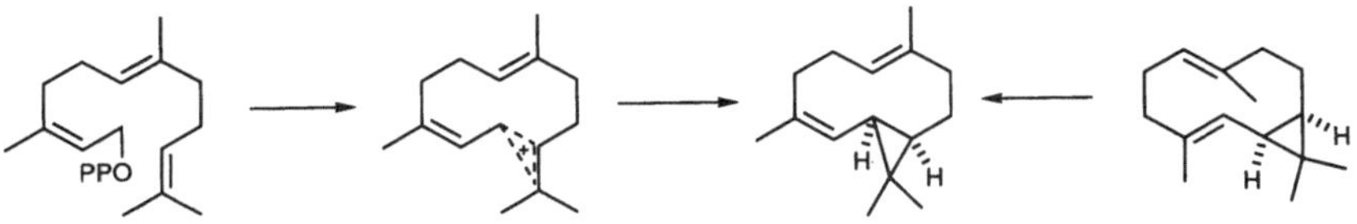

cis-FPP   (141)   (142)   (134)

(143)   (144)

Scheme 5. The biosynthesis of isobicyclogermacrenes

The biosynthesis of the *trans*-fused macrocarpals (129), (130), and (131) from bicyclogermacrene (134) and a benzylic cation has also been explained via a cyclopropyl carbinyl cation (145) as pictured in Scheme 6 (*71*).

Neutralization of the carbocations after cyclization takes place by proton expulsion or reaction with a nucleophile. This process, together

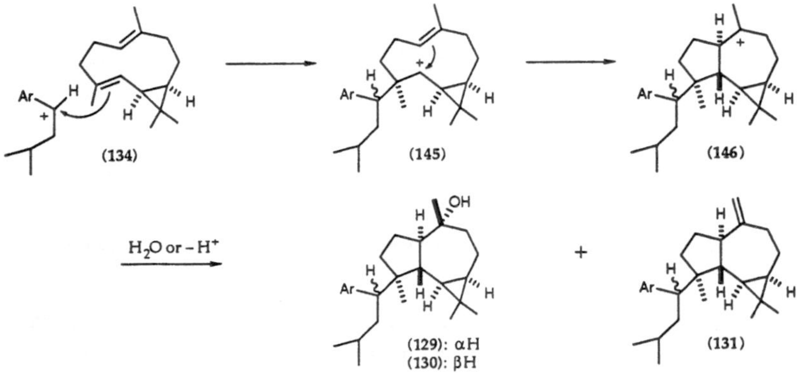

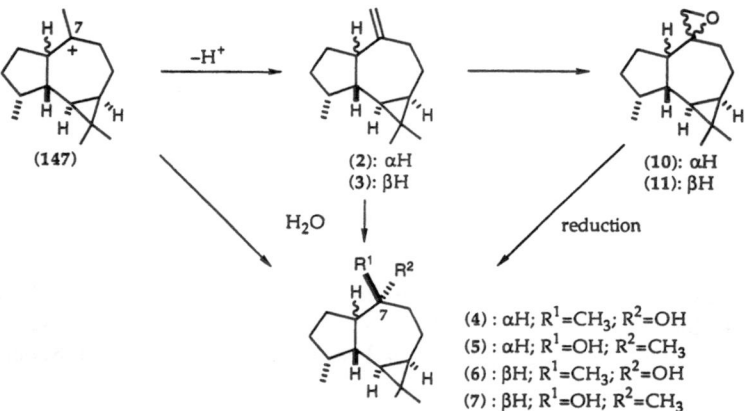

Scheme 6. The biosynthesis of *trans*-fused macrocarpals

with oxidation steps in bicyclogermacrene or aromadendranes, leads to the functional groups found in the naturally occurring aromadendranes. For example, the presence of the tertiary alcohol groups at C7 in (4–7) can be explained by reaction of the C7-cation (147) formed after cyclization with water, or by addition of water to (2) or (3). The reduction of the epoxides (10) and (11), derived from aromadendrene (2) and allo-aromadendrene (3), respectively, has been proposed by Tressl *et al.* for the biosynthesis of (4–7) (*31*) (Scheme 7).

Scheme 7. The biosynthesis of aromadendrane alcohols

The isonitrile groups in natural products found in sponges have been proved to originate from cyanide ions (*89*). The isothiocyanate and formamide groups are subsequently formed from the isonitrile function (*90*).

1,7-Cyclo-, 7,8-seco- (*47*), and 9,10-secoaromadendranes (*55*, *59*) are probably derived from the appropriate cyclization or bond cleavages in the corresponding aromadendranes.

3. Biological Activity

From the aromadendranes listed in Sect. 1 only a few have been shown to exhibit biological activity. This does not necessarily mean that biological activity is rare among aromadendranes. Most compounds have been tested only for one specific biological activity or have not been tested at all.

Many essential oils and other plant extracts used in the fragrance and flavour industry contain aromadendranes (*91*). Most of them are hydrocarbons and/or monohydroxy derivatives. Hop (*Humulus lupulus*), which gives beer its characteristic taste, contains a variety of aromadendranes (*31*, *92*).

Some medicinal plants (*17*, *93*) or plants with known biological activity (*94*) contain aromadendranes, but in most cases it is not known which compounds are responsible for the medicinal properties or biological activities.

Biological activities that have been reported for selected (*ent*)-aromadendranes include antifungal, antibacterial, antiviral (Sect. 3.1), plant growth regulatory (Sect. 3.2), antifeedant, repellent (Sect. 3.3), piscicidal, and cytotoxic activities (Sect. 3.4). Some miscellaneous biological activities are described in Sect. 3.5.

3.1. Antifungal, Antibacterial, and Antiviral Activities

Some aromadendrane mono- and dialcohols have been found to possess fungicidal activity, *e.g.* (−)-ledol (**6**) against *Coriolus ronatus* (*95*) and alloaromadendrane-7β,11β-diol (**41**) against *Cladosporium herbarium* (*19*, *79*). The antifungal properties of (+)-spathulenol (**31**) might be responsible for the observation that it repels leaf cutter ants (*96*) (see also Sect. 3.3). Because of these reported fungicidal activities, the mono- and dihydroxyaromadendranes (**4–7**, **31**, **40** and **41**) have been synthesized and tested for fungitoxicity against *Cladosporium cucumerinum* and *Penicillium italicum* (*97*). The growth inhibition of these fungi was only moderate. (−)-Cyclocolorenone (**27**) inhibits the growth of the fungi *Curvularia lunata*, *Chaetomium cochliodes*, and *Chaetomium spinusum*, but not of *Aspergillus flavus* (*98*). This compound also shows antibacterial activity against several Gram-positive bacteria and, at higher concentration, against Gram-negative bacteria (*98*). The isothiocyanate (**82**) and isonitrile (**84**) isolated from *Acanthella pulcherrima* are, among others, responsible for growth inhibition of the Gram-positive bacteria *Bacillus subtilis* (*43*). Aromadendrane containing essential oils from several plants have been found to possess antibacterial activity (*93b–d*). However, the compounds responsible for this activity have not been identified. Macrocarpals A (**129**), B (**130**), and G (**131**), metabolites from *Eucalyptus macrocarpa*, show antibacterial activity against Gram-positive bacteria such as *Bacillus subtilis* and *Staphylococcus aureus*, but not against Gram-negative bacteria, yeast, or fungi (*72*, *73*).

Macrocarpals (**129–131**) have antiviral properties in that they inhibit the enzyme HIV-reverse transcriptase (*71*). The ledol glycosides, especially arvoside B (**12**), which have been extracted from the aerial parts of *Calendula arvensis*, exhibit antiviral activity against the vesicular stomatitis virus (*27*).

3.2. Plant Growth Regulatory Activity

(+)-Globulol (**4**) shows weak activity against the germination of cress seed (*99*). Alloaromadendrane-7β,11β-diol (**41**) isolated from *Ambrosia peruviana* causes marginal reduction in the growth of cress seeds, but stimulates wood and shoot growth in lettuce at low concentrations (*19*, *79*). (−)-Cyclocolorenone (**27**) shows good growth-inhibitory activity against etiolated wheat coleoptiles. It is also phytotoxic against greenhouse-growth corn, bean, and tobacco plants (*98*).

The *ent*-2,3-secoaromadendrane plagiochiline A (**90**) inhibits the germination of rice and wheat (*53*). The methanol extract from *Plagiochila semidecurrens* inhibits the growth of the leaves and roots of rice seedlings (*59*). The compounds responsible for this plant growth inhibition are the *ent*-2,3-secoaromadendranes ovalifoliene (plagiochiline C) (**96**) (*55*), ovalimethoxy I (**99**) and II (**100**), ovalifolienal (**102**), and ovalifolienalone (**103**). Especially compounds (**96**), (**99**) and (**102**) are very strong growth inhibitors of rice seedlings (*59*).

3.3. Antifeedant and Repellent Activity

Plachiochiline A (**90**), already mentioned in Sect. 3.2, is a very pungent substance found in several *Plagiochila* species. It shows a very strong antifeedant activity against the larvae of the African army worm (*Spodoptera exempta*) (*58*).

(+)-Spathulenol (**31**) is one of the compounds in *Melampodium divaricatum* that has repellent properties against the leaf cutter ant (*Atta cephalotes*) (*96*). It is possible that these ants discriminate against (**31**) because of its antifungal properties (see Sect. 3.1). (−)-Alloaromadendrene (**3**) shows toxic activity against Southeast Asian termites (*Neotermes* spp.) and is probably one of the components in the crude resin of *Dipterocarpus* trees that is responsible for the insecticidal properties of this resin (*100*). The viridiflorol glycosides (**23**) and (**24**) from *Pittosporum tobira* show repellent activity against the blue mussel *Mytilus edulis* (*28*) and might be useful as non-toxic antifouling agents (*101*).

The isonitrile (**77**) and isothiocyanate (**78**), together with other metabolites found in *Axinella* species (sponges), probably secure the sponge against being eaten and overgrown (*102*). A mixture of isonitriles, including (**77**), found in *Axinella* species inhibits the settlement and/or metamorphosis of the larval or juvenile invertebrates *Phidolophora pacifica* (ectoproct), *Salmacina tribranchiata* (polychaete), and *Haliotes refescens* (abalone). Metamorphosis and settlement of *Haliotes refescens* were also inhibited by a mixture of the isothiocyanates, including (**78**), found in *Axinella* species (*102*). Both isonitriles and isothiocyanates inhibited fish feeding when applied to pelleted fish food for goldfish (*Carassius auratus*) and sculpin (*Clinocottus analis*) (*102*). Isonitrile (**77**) and isothiocyanate (**78**), found in the nudibranch *Cadlina luteomarginata*, might have a similar function in the nudibranch as in the sponges (*45*).

3.4. Piscicidal and Other Toxic Activities

Treatment of killie-fish (*Oryzia latipes*) with the pungent plagiochiline A (**90**) at very low concentration kills them within a few hours (*103*). Plachiochiline A also shows cytotoxicity against KB cells (*103*).

The mixture of isonitriles and isothiocynates obtained from *Axinella* species mentioned in Sect. 3.3 shows piscicidal activity against goldfish (*Carassius auratus*) (*102*). The isonitrile (**84**) obtained from *Acanthella acuta* is toxic for guppies (*Lebistis reticulatus*) (*44a,c*). Epipolasinthiourea B (**80**) shows moderate cytotoxicity against L1210 cells (*41*).

The essential oil of *Ledum palustre*, with (+)-ledol (**6**) as a major constituent, has been used as a medicine against various diseases and to provoke abortion. However, ledol is toxic and can cause vomitting and gastroenteritis. On digestion it can cause convulsions, followed by paralysis (*104*).

3.5. Miscellaneous Activities

Plagiochilal B (**108**), isolated from *Plagiochila fruticosa*, causes acceleration of neurite sprouting and enhancement of choline acetyltransferase activity on a neuronal cell culture of fetal rat cerebral hemisphere (*60*). A similar activity has been reported for plagiochilide (**89**) and a mixture of these and other plagiochilines obtained from *Plagiochila fruticosa* might be used as ingredients in nerve cell degeneration reparation agents (*105*).

Macrocarpals A (**129**), B (**130**), and G (**131**) have been shown to inhibit the enzyme aldose reductase from swine (*106*).

The extracts from the aerial parts of *Calendula arvensis*, containing ledol glycosides (**12–17**) (*27*), show anti-inflammatory activity (*93a*). Euglobal V (**128**) causes strong granulation-inhibition in the fertile egg test, indicating anti-inflammatory activity (*70a*).

4. Total Syntheses of Aromadendranes

The available literature on the syntheses and the chemistry of aromadendranes can be divided into two parts: total syntheses of aromadendranes (Sect. 4) and syntheses with natural aromadendranes as starting material (Sect. 5).

Most total syntheses of aromadendranes have been based on the construction of a hydronaphthalene precursor that is subsequently rearranged to a hydroazulene ring system. The introduction of substituents in a stereoselective way is more easily achieved in the conformationally well understood hydronaphthalene system than in the hydroazulene system with its flexible cycloheptane ring. This makes the approach via hydronaphthalene systems attractive and reliable.

Several types of rearrangement from hydronaphthalene to hydroazulene skeletons have been described (*107*). The well known total synthesis of (−)-aromadendrene (**57**) by BÜCHI et al. (*3*) uses a pinacoltype rearrangement (Scheme 8). In this synthesis the hydronaphthalene skeleton was constructed via a Diels-Alder condensation of diene (**149**),

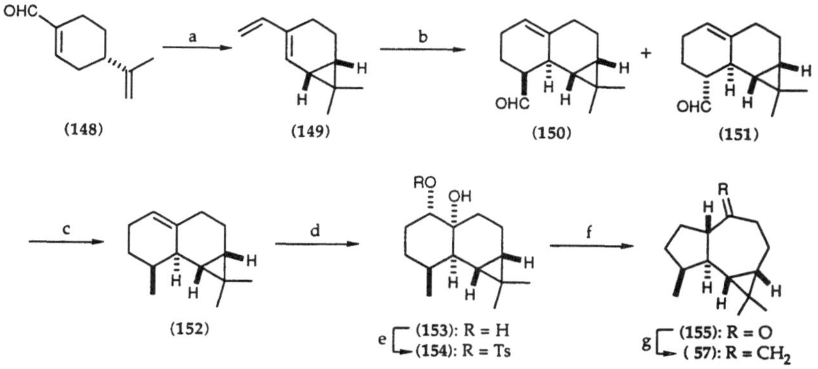

(a) HBr; KOtBu, HOtAm; Ph$_3$P = CH$_2$; (b) acrolein, 100 °C; (c) LiAlH$_4$; MsCl; LiAlH$_4$;
(d) OsO$_4$; (e) TsCl; (f) Al$_2$O$_3$; (g) Ph$_3$P = CH$_2$.

Scheme 8. The total synthesis of (−)-aromadendrene

which was prepared from (+)-perillaldehyde (**148**), with acrolein. This cycloaddition gave a 5:1 mixture of the aldehydes (**150**) and (**151**), respectively.

Reduction of the aldehyde group in (**150**) to the methyl group by standard procedures followed by oxidation of the double bond with OsO$_4$ gave a single diol (**153**) which was selectively tosylated to (**154**). When (**154**) was treated with activated alumina in CH$_3$Cl, it readily underwent a pinacol rearrangement to give (−)-apoaromadendrone (**155**). Ketone (**155**) was subsequently converted into (−)-aromadendrene (**57**) via a Wittig reaction.

SURBURG and MONDON (*108*) followed the same reaction sequence in their synthesis of (−)-spathulenol (**64**) (Scheme 9). By using acroleic acid as dienophile, instead of acrolein, in the Diels-Alder condensation with diene (**149**), hydronaphthalene (**156**) and its C11-epimer were obtained again in a 5:1 ratio, respectively.

Following the procedure of BÜCHI et al. (*3*), the rearranged methyl ester (**158**) was prepared. After oxidative decarboxylation of its methylthio derivative (**159**), ketone (**160**) was obtained in low yield. Grignard addition with MeMgI, followed by hydrolysis of the acetal and a Wittig reaction afforded the alcohols (−)-spathulenol (**64**) and 11-epispathulenol (**161**) in a 1:1 ratio.

JENNISKENS et al. (*79*) have developed a new route to synthesize *cis*-fused hydroazulene systems via a base-induced and -directed rearrangement of substituted *trans*-perhydronaphthalene-1,4-diol monosulfonate

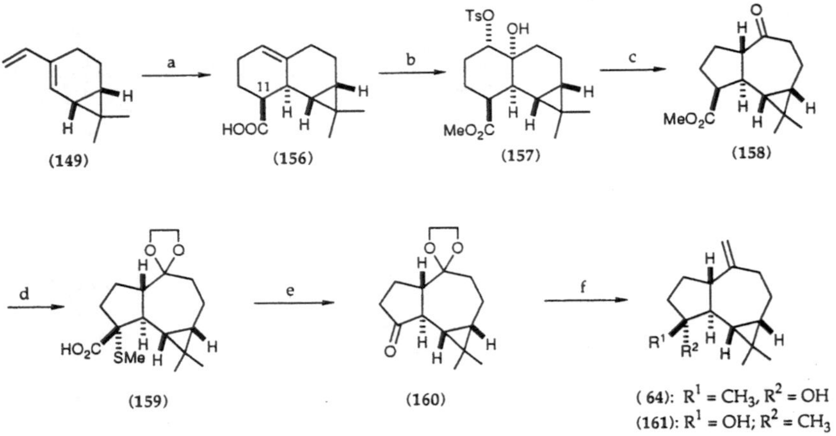

(149)   (156)   (157)   (158)

(159)   (160)   (64): R^1 = CH_3, R^2 = OH
                              (161): R^1 = OH; R^2 = CH_3

(a) acroleic acid, 100 °C; (b) OsO_4; CH_2N_2; TsCl; (c) KOtBu, HOtAm; (d) ethylene glycol, pTsOH, benzene, LDA, Me_2S_2; KOH, ethylene glycol; (e) NCS, MeOH; aq. HCl; (f) MeMgI; aq. H_2SO_4, CH_2Cl_2; Ph_3P = CH_2.

Scheme 9. The total synthesis of (−)-spathulenol

esters. This method has been applied to the total synthesis of (±)-alloaromadendrane-7β,11α-diol (**168**) supposedly isolated from *Ambrosia peruviana* (*19*) (Scheme 10). Starting from the readily available compound

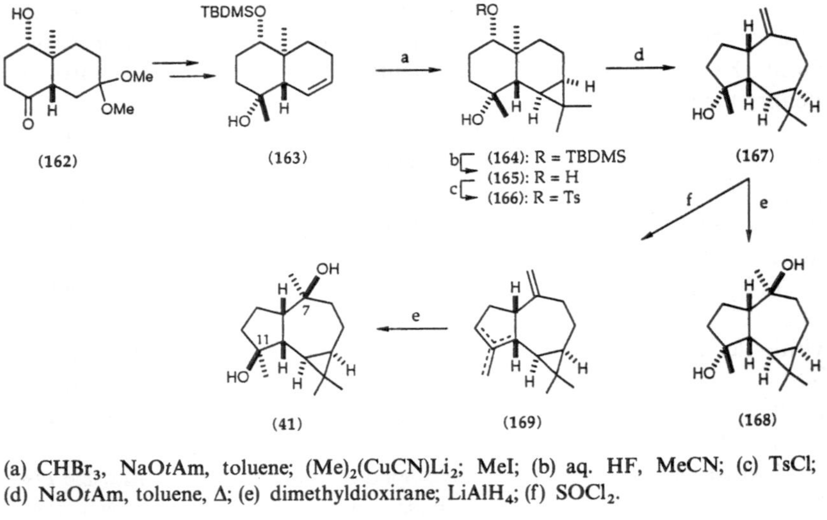

(162)   (163)   b⌈ (164): R = TBDMS   (167)
                              c⌊ (165): R = H
                                 (166): R = Ts

(41)   (169)   (168)

(a) $CHBr_3$, NaOtAm, toluene; $(Me)_2(CuCN)Li_2$; MeI; (b) aq. HF, MeCN; (c) TsCl; (d) NaOtAm, toluene, Δ; (e) dimethyldioxirane; $LiAlH_4$; (f) $SOCl_2$.

Scheme 10. The total synthesis of (±)-alloaromadendrane-7β,11β-diol (**41**)

(162), the *trans*-fused hydronaphthalene-1,4-diol monosulfonate ester **(166)** was prepared. Base-induced rearrangement of **(166)** with Na *t*-amylate in refluxing toluene gave, via selective intramolecular deprotonation, the alloaromadendrene derivative **(167)** in 70% yield.

Epoxidation and subsequent reduction of **(167)** gave $(\pm)$-allo-aromadendrane-7β,11α-diol **(168)**. However, its spectral data did not agree with those reported for the natural product. On the other hand, the spectral data of the epimeric 7β,11β-diol **(41)**, prepared from **(167)** via a dehydration, epoxidation, and reduction sequence, agreed very well with those of the natural product. Consequently, the natural product isolated from *A. peruviana* possesses the stereochemistry shown in structure **(41)** and not the one proposed in structure **(168)**.

Another approach to obtain hydroazulene products from hydro-naphthalene precursors is the photochemical rearrangement of cross-conjugated hydronaphthalene dienones. α-Santonin **(170)** can be rearranged to the O-acetylisophotosantonic lactone **(171)** upon irradiation in aqueous AcOH (*109*). This lactone has been transformed into 8-epicyclocolorenone **(177)** by Büchi *et al.* (*8*) (Scheme 11).

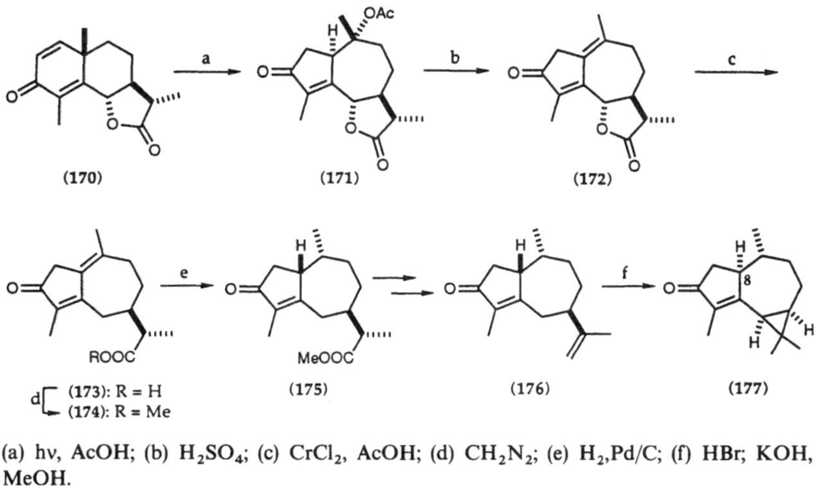

(a) hv, AcOH; (b) H₂SO₄; (c) CrCl₂, AcOH; (d) CH₂N₂; (e) H₂,Pd/C; (f) HBr; KOH, MeOH.

Scheme 11. The total synthesis of 8-epicyclocolorenone

Treatment of the acetate **(171)** with concentrated H₂SO₄ afforded the dienone lactone **(172)**. Treatment of **(172)** with chromous chloride in AcOH gave the carboxylic acid **(173)**. Partial hydrogenation of the methyl ester **(174)** with H₂ and Pd/C gave **(175)**, which could be converted into the olefin (176) in a few steps. Construction of the cyclopropane ring in (177) proceeded via treatment of the tertiary bromide, prepared from (176), with methanolic KOH. The basic conditions used in the last step caused the C8-proton to epimerize to the α-position.

STREITH and BLIND used the hydronaphthalene (179), obtained by oxidation of (−)-7-epicyperone (178) with DDQ, for their synthesis of cyclocolorenone derivatives (110) (Scheme 12). Photochemical rearrangement of (179) in AcOH gave a mixture of the hydroazulenes (180) and (181). Cyclopropane ring closure, as described above, gave the cyclocolorenone derivatives (182), (183), and (184).

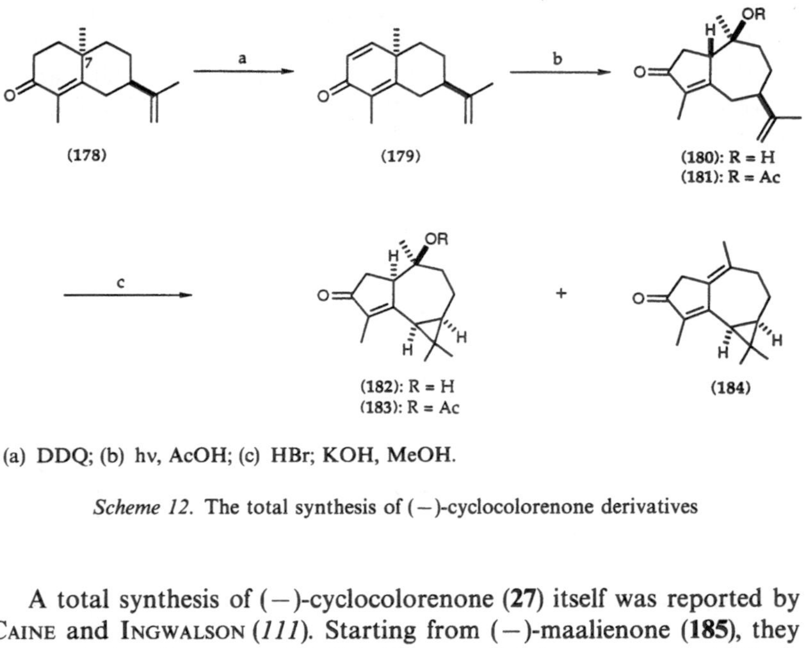

(a) DDQ; (b) hv, AcOH; (c) HBr; KOH, MeOH.

Scheme 12. The total synthesis of (−)-cyclocolorenone derivatives

A total synthesis of (−)-cyclocolorenone (27) itself was reported by CAINE and INGWALSON (111). Starting from (−)-maalienone (185), they prepared the dienone acid (186) (Scheme 13). Irradiation of (186) in dioxane gave the trienone (188) in 60% yield. Subsequent reduction of the exocyclic double bond with H_2 and Pd/C afforded (−)-cyclocolorenone (27).

In contrast to the dienone acid (186) the dienone (187) was photochemically stable. However, its C7-epimer (189) appeared to be photolabile and was used as a precursor in the synthesis of 11-epiglobulol (191) and the olefinic aromadendranes 11-epiaromadendrene (192) and 11-epiledene (193) (112) (Scheme 14).

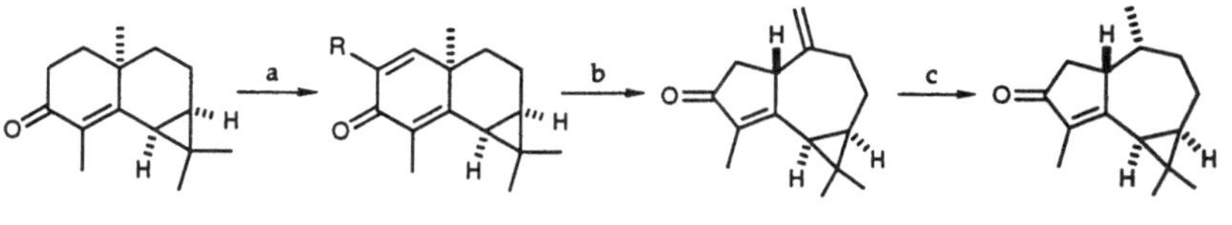

(185)   (186): R = COOH   (188)   (27)
         (187): R = H

(a) HCO$_2$Et, NaOAc; DDQ; CrO$_3$; (b) hv, dioxane; (c) H$_2$, Pd/C.

Scheme 13. The total synthesis of (−)-cyclocolorenone

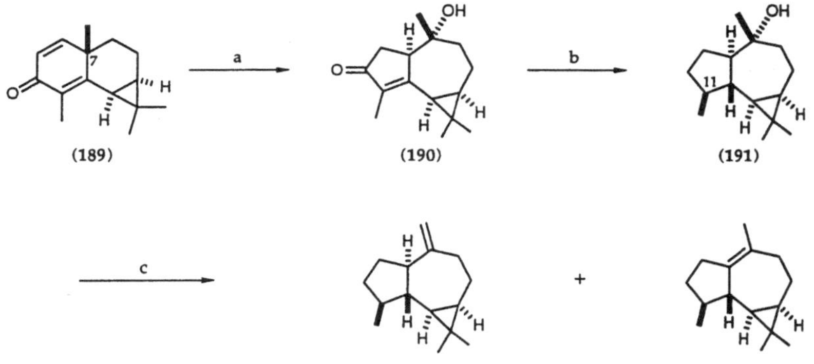

(189)   (190)   (191)

(192)   +   (193)

(a) hv, HOAc/H$_2$O; (b) Li, NH$_3$; Wolff-Kishner; (c) SOCl$_2$; NaOAc, HOAc.

Scheme 14. The synthesis of 11-epiaromadendrene and 11-epiledene

11-Epiledene (193) has also been synthesized by an approach that does not involve a hydronaphthalene precursor (*113*). Starting from (+)-2-carene (194) the phenylthio(trimethylsilyl)cyclopropyl acetal (195) was prepared (Scheme 15). Reductive lithiation with lithium-1-(dimethylamino)naphthalenide (LDMAN) gave the α-lithiosilane (196) which was condensed with crotonaldehyde to give the alcohols (197) and (198) in a 7:1 ratio, respectively. The alcohols (197) and (198) underwent a Peterson olefination upon treatment with KH to give the allylidenecyclopropanes (199) and (200), respectively. Upon thermolysis, both (199) and (200) gave a ca. 1:1 mixture of the aromadendrane dienes (201) and (202). Hydrogenation of (201) using Wilkinson's catalyst gave 11-epiledene (193).

A second approach to aromadendrenes which does not use a hydronaphthalene precursor was described by NARANG and DUTTA (*114*)

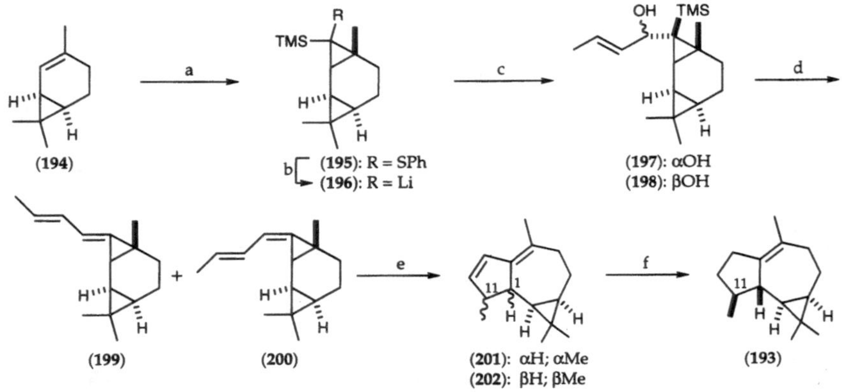

(a) PhSCHCl$_2$, KOH; sBuLi; TMSCl; (b) LDMAN; (c) crotonaldehyde; H$_2$O; (d) KH;
(e) thermolysis 190 °C; (f) H$_2$, (PPh$_3$)$_3$RhCl.

Scheme 15. The total synthesis of 11-epiledene

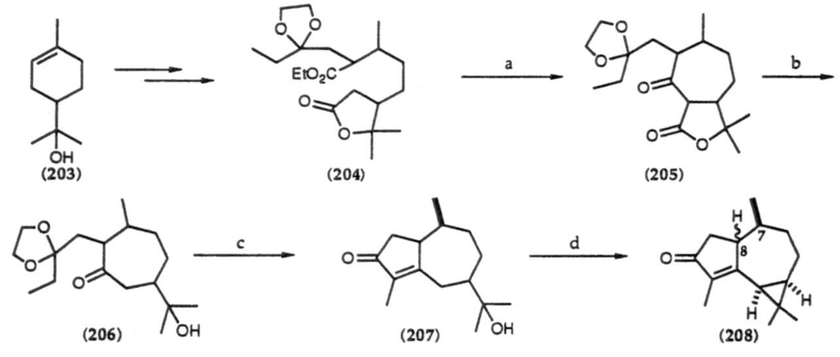

(a) KOtBu, xylene, Δ; (b) KOH, H$_2$O; (c) pTsOH; KOH, H$_2$O; (d) HBr; KOH, MeOH.

Scheme 16. The synthesis of a cyclocolorenone epimer

(Scheme 16). Starting from ($\pm$)-terpineol (203), the 7-membered ring precursor (206) was synthesized in two steps from (204). Hydrolysis of the acetal group followed by treatment with aqueous KOH afforded the hydroazulenic product (207). The cyclopropane ring was constructed according to standard procedures to give in low yield compound (208), a C7-epimer of cyclocolorenone (27) with unknown stereochemistry at C8.

A biomimetic strategy towards the synthesis of aromadendranes makes use of a transannular cyclization of cyclodecane derivatives. Two examples of this approach have already been described in Sect. 2 in the synthesis of ledene (**43**) (*82*) and guayulin C (**33**) and D (**34**) (*18*) from bicyclogermacrene precursors.

In a synthesis of ($\pm$)-globulol (**4**) MARSHALL and RUTH prepared the hydroazulenic compound (**210**) by a stereoselective solvolytic cyclization of the cyclodecadienol derivative (**209**) (*115*) (Scheme 17). Compound (**210**) could be converted into ($\pm$)-globulol (**4**) upon reaction with dibromocarbene, followed by treatment of the resulting dibromocyclopropane derivative (**211**) with lithium dimethylcuprate.

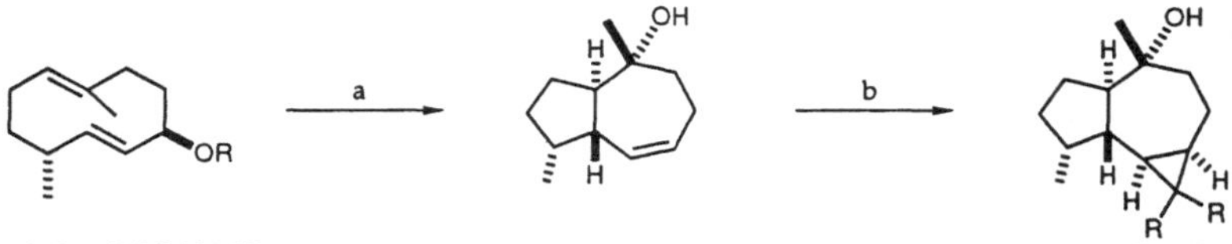

(**209**): R = $COC_6H_4NO_2$-*p*   (**210**) $c\left[\begin{array}{l}(\mathbf{211}): R = Br \\ (\mathbf{4}): R = Me\end{array}\right.$

(a) $NaHCO_3$, H_2O; (b) $PhHgCBr_3$, benzene; (c) $LiMe_2Cu$.

Scheme 17. The total synthesis of ($\pm$)-globulol

The 7,8-secoaromadendrane ($-$)-taylorione (**110**) has been synthesized by NAKAYAMA et al., starting from Δ^3-carene (**212**) (*116*) (Scheme 18). Ozonolysis of (**212**) followed by oxidation and esterification gave product (**213**) which was transformed to product (**214**) in several steps. Removal of both protective groups with aqueous HCl in acetone gave the γ-keto-aldehyde (**215**). Treatment of (**215**) with methanolic NaOH afforded the cyclisized product (**216**) which was converted into ($-$)-taylorione (**110**) via a Wittig reaction and subsequent oxidation of the alcohol group.

PATTENDEN and WHYBROW tried to synthesize the taylorione skeleton via the photosensitive Z- and E-cyclopentenones (**217**) and (**218**) (*117*) (Scheme 19). Irradiation of the E-cyclopentenone (**218**) gave via a di-π-methane rearrangement the cyclopropane (**219**), which was converted into *trans*-deoxytaylorione (**220**) via a Wittig reaction. Irradiation of the Z-cyclopentenone (**217**) also led to (**219**), presumably via (**218**), which was produced by Z-E photoisomerization.

The bicyclic enone (**221**), available in both racemic and chiral forms (*118*), might be a versatile intermediate for the synthesis of (seco)-aromadendranes. It has been used as starting material for the synthesis of the 9,10-secoaromadendrane ($+$)-hanegokedial (**107**) (*119*) (Scheme 20). Treatment of ($-$)-(**221**) with bis(1,1-diethoxy-2-propenyl) lithium cuprate

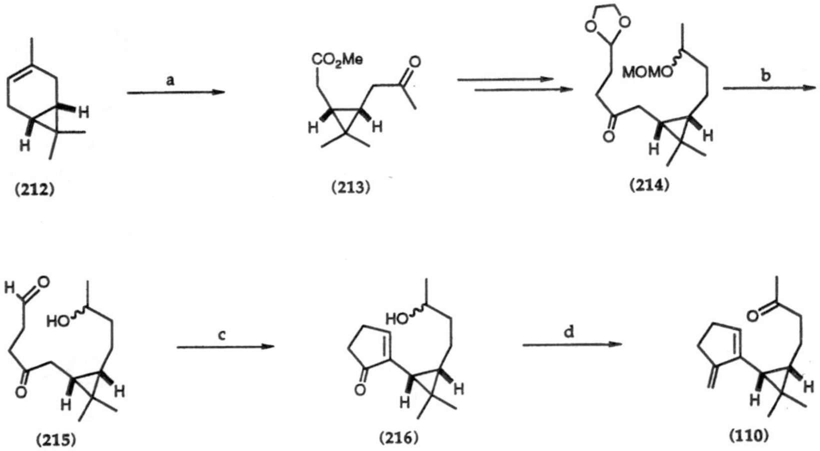

(a) O_3; CrO_3; MeOH, HCl; (b) HCl, H_2O, acetone; (c) NaOH, MeOH; (d) $Ph_3P = CH_2$; DMSO, benzene, DCC, pyridine-$OCOCF_3$.

Scheme 18. The total synthesis of (−)-taylorione

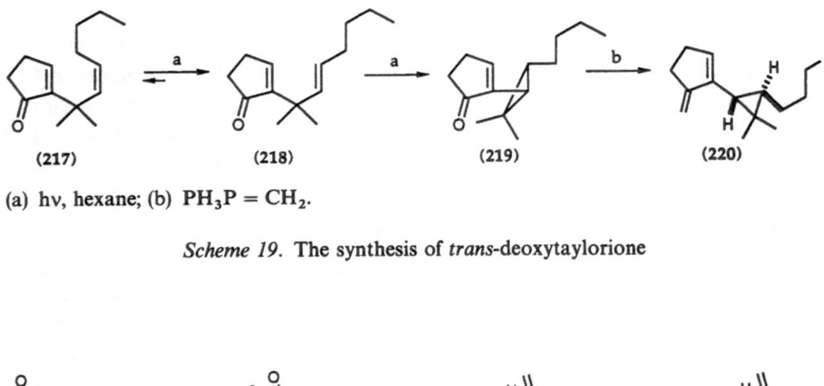

(a) hv, hexane; (b) $PH_3P = CH_2$.

Scheme 19. The synthesis of *trans*-deoxytaylorione

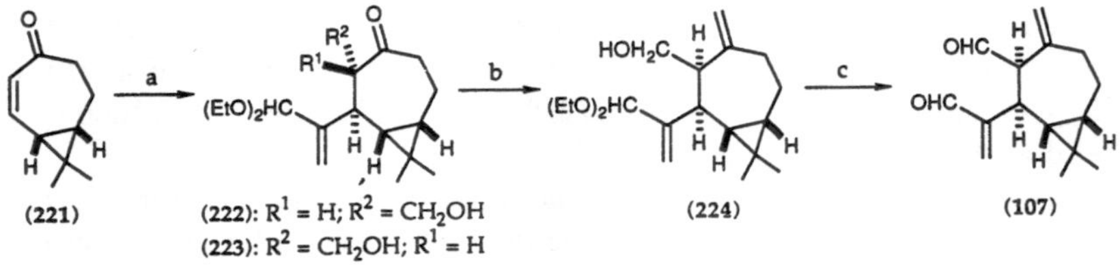

(a) Li(1,1-$(EtO)_2$-2-propenyl)Cu; H_2CO; (b) $Ph_3P = CH_2$; (c) Collin's reagent; H_3O^+.

Scheme 20. The total synthesis of (+)-hanegokedial

in ether and quenching of the resultant enolate with formaldehyde gave the isomeric alcohols (222) and (223) in a 2:1 ratio, respectively. The minor alcohol (223) was converted into (224) via a Wittig reaction. Oxidation of (224), followed by hydrolysis of the acetal gave (+)-hanegokedial (107). The difference between the optical rotations of natural (−)-hanegokedial (55) and the synthesized (+)-hanegokedial, respectively −10.4° and +0.5°, is probably due to the presence of impurities formed during storage of the unstable hanegokedial.

5. Aromadendranes as Starting Materials in the Syntheses of Sesquiterpenes

Several aromadendranes have been synthesized from more or less readily available aromadendranes. Usually, this has been done to correlate the structure and stereochemistry of newly isolated aromadendranes with those of known ones.

Epoxidation of aromadendrene (2) or alloaromadendrene (3), followed by reduction gave the tertiary alcohols globulol (4) and epiglobulol (5), or ledol (6) and viridiflorol (7), respectively (31, 74, 120) (Scheme 21). Epiglobulol could also be obtained by reaction of MeLi or MeMgI with

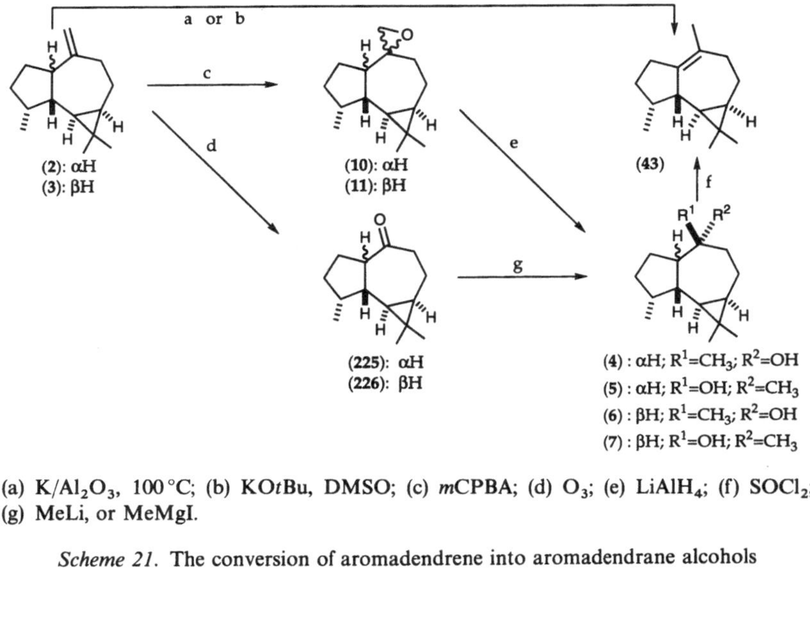

(a) K/Al$_2$O$_3$, 100 °C; (b) KOtBu, DMSO; (c) mCPBA; (d) O$_3$; (e) LiAlH$_4$; (f) SOCl$_2$; (g) MeLi, or MeMgI.

Scheme 21. The conversion of aromadendrene into aromadendrane alcohols

(+)-apoaromadendrone (**225**), prepared from (+)-aromadendrene by ozonolysis (*121*). In a similar way ledol could be prepared from allo-aromadendrone (**226**) (*121*). Dehydration of the tertiary alcohols (**4–7**) gave mainly (+)-ledol (**43**) (*74*). Isomerization of (+)-aromadendrene with K/Al$_2$O$_3$ at 100 °C gave (+)-ledene in 40% yield (*122*). Isomerization of (+)-aromadendrene with KOtBu in DMSO gave (+)-ledene in 80% yield (*123*).

Large quantities of the crystallizable (+)-apoaromadendrone (**225**) could be obtained by ozonolysis of a distillation tail fraction of *Eucalyptus* oil, containing 55–70% of (+)-aromadendrene (**2**) and 10–15% of alloaromadendrene (**3**) (*124, 125*). The *cis*-fused alloaromadendrone (**226**) could be obtained from the *trans*-fused **225** via a selective protonation of the thermodynamic enol trimethylsilylether (**227**) (Scheme 22) (*97*). In this way a straightforward synthesis of ledol (**6**) could be accomplished in high overall yield (*128*).

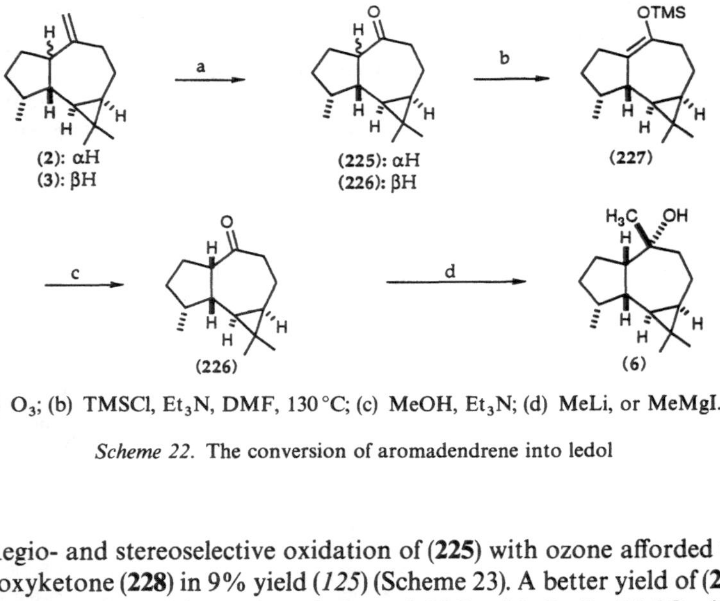

(a) O$_3$; (b) TMSCl, Et$_3$N, DMF, 130 °C; (c) MeOH, Et$_3$N; (d) MeLi, or MeMgI.

Scheme 22. The conversion of aromadendrene into ledol

Regio- and stereoselective oxidation of (**225**) with ozone afforded the hydroxyketone (**228**) in 9% yield (*125*) (Scheme 23). A better yield of (**228**) (36%) was obtained by hydroxylation of (**225**) with RuO$_2$/NaIO$_4$ (*97*). Hydroxyketone (**228**) was converted into (+)-spathulenol (**31**) by a Wittig reaction or Peterson olefination. Repeated epimerization of (**228**) afforded the *cis*-fused (**229**) (*97*). The *cis*- and *trans*-hydroxyketones (**228**) and (**229**), and (+)-spathulenol (**31**) (*15, 35*) have been used to synthesize the *cis*-diols (**41**) and (**230**), and the *trans*-diols (**40**) and (**231**) by similar

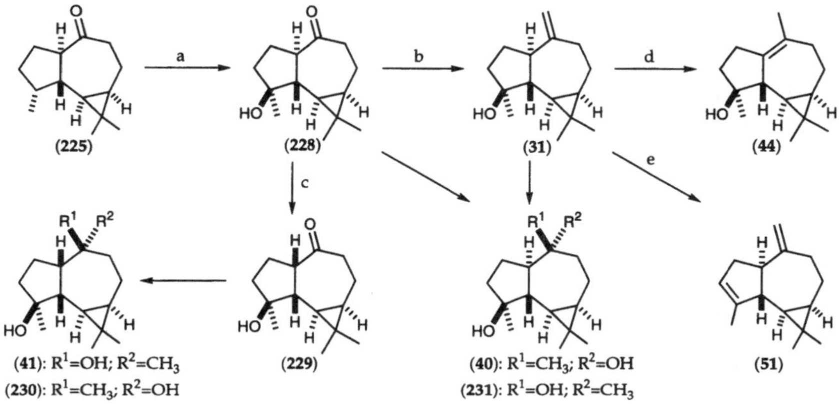

(a) O_3 or RuO_2, $NaIO_4$; (b) $Ph_3P = CH_2$ or $TMSCH_2MgCl$; KH; (c) $NaOCH_3$, CH_3OH;
(d) pTsOH; (e) $SOCl_2$.

Scheme 23. The oxidation of alloaromadendrone to aromadendrane diols

procedures as described for the mono-alcohols (**4–7**) (*97*) (vide supra).
Spathulenol (**31**) has also been used as starting material in the synthesis of
isospathulenol (**44**) (*32*) and β-spathulene (**51**) (*12*).

(+)-Aromadendrene (**2**) has also been used for the synthesis of
sesquiterpenes with a maaliane (*126*), guaiane (*123, 127*), humulane (*128*)
or cadinane (*128*) skeleton.

Epoxidation of the enol trimethylsilylether (**227**), prepared from (**2**)
(*97*) (vide supra), gave the hydroazulene α-ketol (**232**) (Scheme 24).
Starting from (**232**), two different routes to hydronaphthalene compounds with a maaliane skeleton have been developed (*126*).

In the first route, stirring of (**232**) with Al_2O_3 gave α-ketol (**233**).
Treatment of (**233**) with lithium in NH_3, followed by addition of MeI,
gave the *cis*-fused maaliane ketone (**234**).

In the second route the β-silyloxy epoxide (**235**) could be rearranged
in high yield with $TiCl_4$ to the *trans*-fused maaliane compound (**236**). The
epoxide (**235**) was synthesized from (**232**) via epoxidation of the naturally
occurring aromadendrene alcohol (**9**). The rearranged compound (**236**)
was reduced in two steps to remove both oxygen functions. Hydroxylation of C11 with $RuO_2/NaIO_4$ gave the naturally occurring (+)-
maaliol (**237**) in moderate yield (*126*).

Selective, acid-catalyzed cleavage of the C3–C4 bond of the cyclopropane ring in (**225**) (and (**226**)) gave (**238**) in high yield (*125*) (Scheme 25).
Ozonolysis of (**238**) afforded the keto alcohol (**239**) which is a suitable

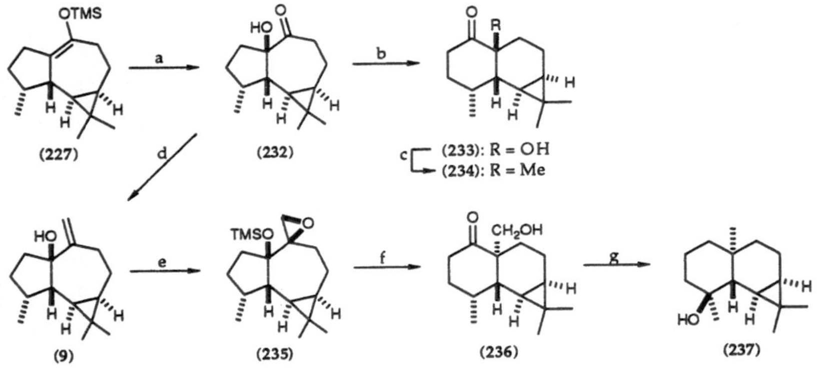

(a) dimethyldioxirane; SiO$_2$; (b) Al$_2$O$_3$; (c) Li, NH$_3$, tBuOH; MeI; (d) TMSCl, HMDS; Ph$_3$P = CH$_2$; TBAF; (e) tBuOOH, VO(Acac)$_2$; TMSCl, HMDS; (f) TiCl$_4$; (g) tosylhydrazine, NaBH$_3$CN, ZnCl$_2$, CH$_3$OH, Δ; nBuLi, bis(dimethylamino)-chloro-phosphoramidate; Li, EtNH$_2$, tBuOH; RuO$_2$, NaIO$_4$.

Scheme 24. The synthesis of (+)-maaliol

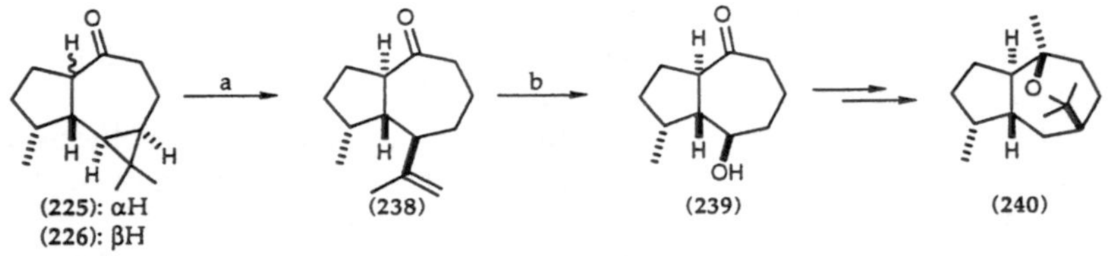

(a) HCl, EtOH, Δ; (b) O$_3$, MeOH; Ac$_2$O, Et$_3$N, DMAP; NaOCH$_3$, CH$_3$OH.

Scheme 25. The synthesis of (−)-kessane

chiral intermediate for the syntheses of guaianes. This has been demonstrated in the synthesis of (−)-kessane (**240**), which proceeded in a 9 steps reaction sequence in an overall yield of 43% from **239** (*127*).

Another guaiane sesquiterpene, (+)-γ-gurjunene (**241**), has been synthesized by pyrolysis of isoledene (**240**) (*123*) (Scheme 26). The latter product was obtained by treatment of pure (**2**) (*122*) or a crude mixture of (**2**) and alloaromadendrene (**3**) (*128*) with K/Al$_2$O$_3$ at room temperature. Both (**2**) and (**3**) are quantitatively converted into isoledene (**240**). Oxidative cleavage of the central double bond in (**240**) produced the thermolabile bicyclogermacrane-1,8-dione (**242**) (*128*). Thermal rearrangement of (**242**) gave as a result of a homo [1,5] hydrogen shift at relatively low temperature (refluxing dioxane) the humulane compound

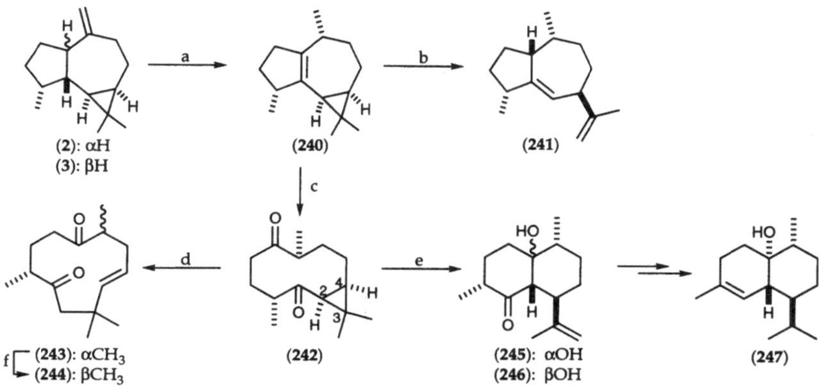

(2): αH
(3): βH

(240)

(241)

f [(243): αCH₃
  [(244): βCH₃

(242)

(245): αOH
(246): βOH

(247)

(a) K/Al₂O₃, room temperature; (b) 450 °C; (c) RuO₂, NaIO₄; (d) dioxane, Δ; (e) FVP, 500 °C to 700 °C; (f) NaOCH₃, CH₃OH.

Scheme 26. The synthesis of humulenedione and (−)-cubenol

(243) and at higher temperature (flash vacuum pyrolysis, 500 °C and up) the products (245) and (246), both with a cadinane skeleton. Epimerization of (243) gave the naturally ocurring humulenedione (244). Starting from (245), the naturally occurring (−)-cubenol (247) was synthesized in a 4 steps reaction sequence (*128*).

Many aromadendranes are oxidized derivatives of more common aromadendranes. Several of these oxidized derivatives have been prepared by allylic or microbial oxidation of abundant aromadendranes (Scheme 27). Thus squamulosone (45) has been synthesized by allylic oxidation of (+)-ledene (43) with SeO₂ (*29*). 8-Hydroxy-alloaromadendrene (9) has been prepared in a similar way from alloaromadendrene (3) (*34*). Microbial hydroxylation of (3) with *Mycobacterium smegmatis* also gave (9) (*129*). (+)-Aromadendrene (2) and globulol (4) were also subjected to oxidation by this microorganism. In this way the two hydroxylated products (248) and (249) were obtained from (2), the acid (250) was obtained from (4) (*130*).

Fermentation of (4) with *Diplodia gossypina* gave the three hydroxylated products (251), (252), and (253) (*130*). The same products were formed by fermentation with *Bacillus megaterium*, together with three other products, (254), (255), and (256) (*130*). The yields of these biotransformations were very low, with the exception of the formation of (250) from (4) which proceeded in 46% yield (*130*). Autooxidation of (+)-palustrol (63) afforded the diol (257) (*37a*). (−)-Cyclocolorenone (27) was metabolized by rabbits to the oxidized products (258) and (259) (*131*).

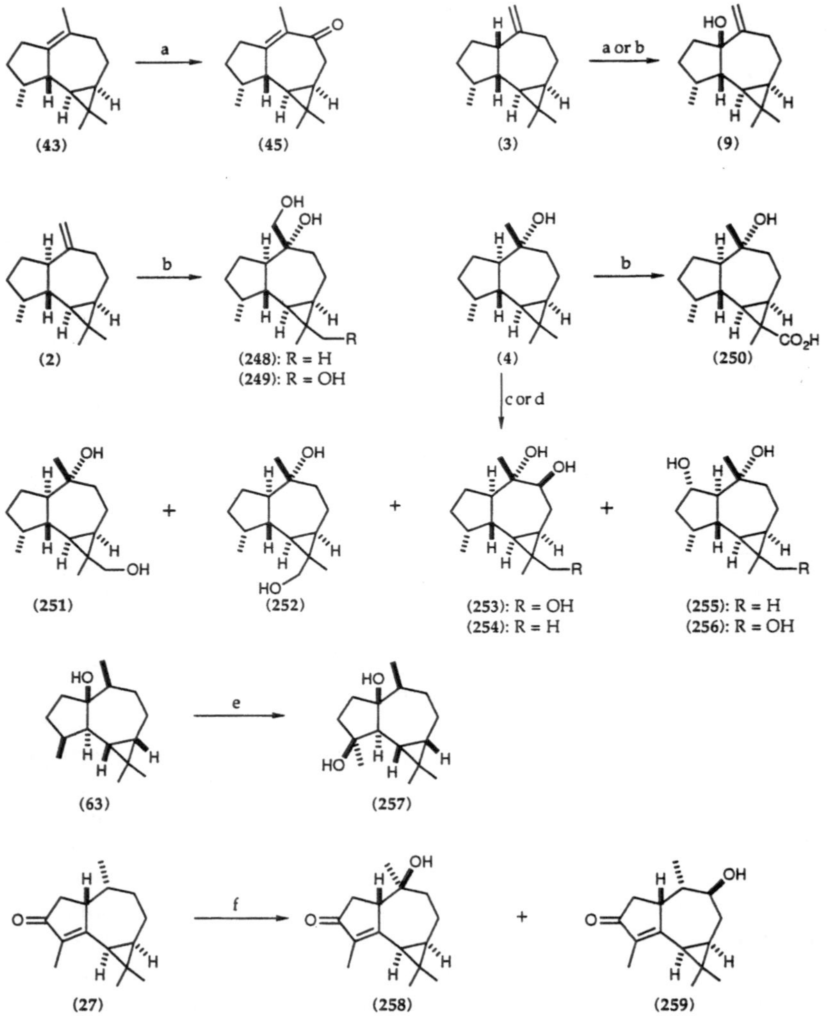

(a) SeO$_2$; (b) *M. smegmatis*; (c) *D. gossypina*; (d) *B. megaterium*; (e) O$_2$; (f) rabbit.

Scheme 27. Allylic and microbial oxidation of aromadendranes

(−)-Cyclocolorenone (**27**) has been prepared from (−)-α-gurjunene (**25**) to establish the stereochemistry of α-gurjunene (Scheme 28) (*5a, 6*). Treatment of (**25**) with sodium peroxide gave (**27**) in very low yield (<2%) (*5a*). Oxidative hydroboration of (**25**) gave the alcohol (**260**) in

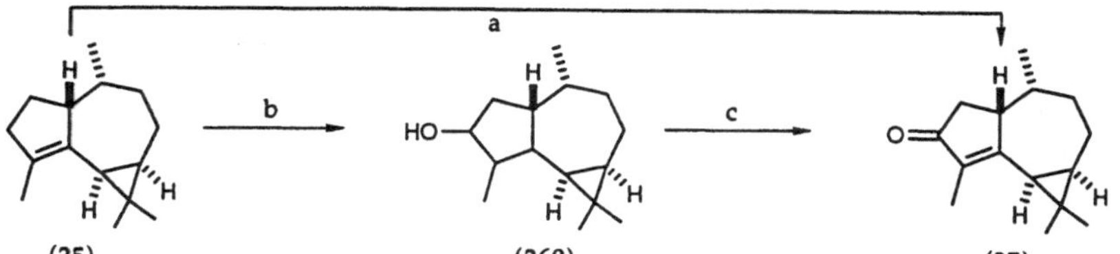

(a) Na_2O_2; (b) BH_3; NaOH, H_2O_2; (c) CrO_3; phenyltrimethylammonium tribromide; Li_2CO_3, LiCl, DMF.

Scheme 28. The conversion of $(-)$-α-gurjunene into $(-)$-cyclocolorenone

about 25% yield. Oxidation of (260) followed by bromination and dehydrobromination gave (27) in about 50% overall yield (6).

α-Gurjunene (25) has also been used as starting material in the synthesis of various other types of sesquiterpenes, mainly guaianes (Scheme 29). Treatment of (25) with acid provided a 3:2 mixture of the guaiane-type products (261) and (262), respectively (132). Guaiadiene

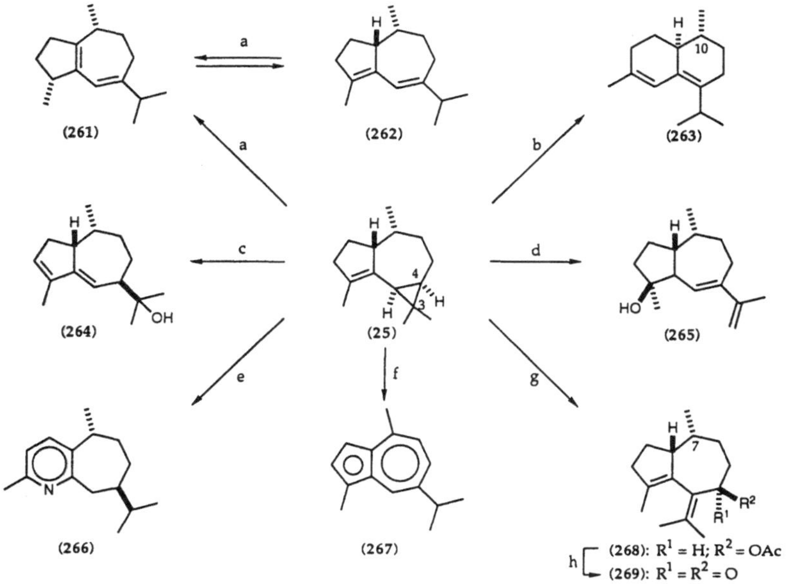

(a) pTsOH, AcOH; (b) H_2SO_4; (c) SeO_2; (d) mCPBA; (e) O_3; $HONH_2$-HCl; (f) S, Δ;
(g) hv, O_2; $NaBH_4$; oxalic acid, Ac_2O, HOAc; (h) $LiAlH_4$, CrO_3-pyridine.

Scheme 29. The conversion of $(-)$-α-gurjunene into guaianes

(261) has been isolated from tolu balsam (133). When strongly acidic conditions (conc. H_2SO_4) were used, (25) was rearranged to 10-epizonarene (263) in 50% yield (134). The insecticidal dienols (264) and (265) (γ-gurjunenol) isolated from *Dipterocarpus kerii* tree resins have been prepared from (25) by oxidation with SeO_2 and *m*CPBA, respectively (135). The alkaloid epiguaipyridine (266) isolated from patchouli oil was synthesized from (25) in two steps (10). Oxidative ring opening of (25) gave a diketone which, after treatment with hydroxylamine hydrochloride, gave (266) in low yield. Dehydrogenation of (25) (5a), as well as of (+)-aromadendrene (2) (136), with sulfur, gave the blue S-guaiazulene (267).

Cleavage of the C3-C4 bond of the cyclopropanering in (25) could be achieved by sensitized photooxygenation of (25) (137). Reduction of the crude photooxygenized mixture with $NaBH_4$, followed by treatment with oxalic acid in $Ac_2O/HOAc$ gave 10-epizieryl acetate (268) in low yield. After reduction of the acetate and subsequent oxidation of the alcohol function, epizierone (269), a C7-epimer of the naturally occurring zierone, was obtained.

The 1,7-cycloaromadendrane (−)-dihydromylione A (117) has been synthesized from (−)-myliol (114) by hydrogenation of the double bond followed by oxidation of the hydroxyl group (64) (Scheme 30). Treatment of (114) with methanolic HCl gave (117) directly.

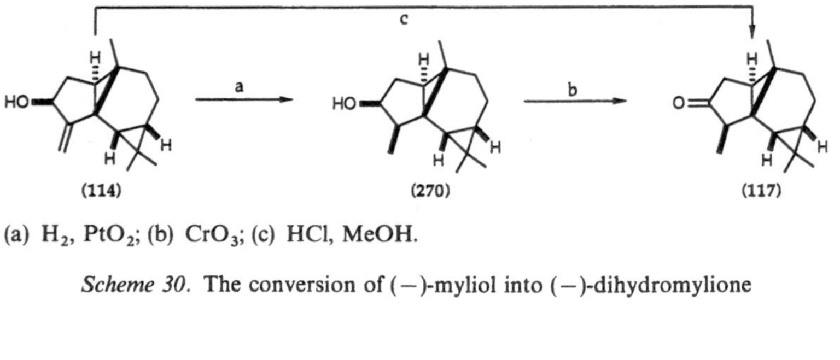

(a) H_2, PtO_2; (b) CrO_3; (c) HCl, MeOH.

Scheme 30. The conversion of (−)-myliol into (−)-dihydromylione

References

1. SMITH, H.G.: Note on the Sesquiterpene of *Eucalyptus* Oils. Proc. Roy. Soc. N.S. Wales, **35**, 124 (1901).
2. BIRCH, A.J., and F.N. LAHEY: The Structure of Aromadendrene, I. Austral. J. Chem., **6**, 379 (1953).
3. (a) BÜCHI, G., W. HOFHEINZ, and J.V. PAUKSTELIS: Total Synthesis of (−)-Aromadendrene. J. Amer. Chem. Soc., **88**, 4113 (1966). (b) BÜCHI, G., W. HOFHEINZ, and J.V. PAUKSTELIS: Synthesis of (−)-Aromadendrene and Related Sesquiterpenes. J. Amer. Chem. Soc., **91**, 6473 (1969).

4. (a) PALMADE, M., and G. OURISSON: La Structure de l'α-Gurjunene (Note Préliminaire). Bull. Soc. Chim. France, 886 (1958). (b) TREIBS, W., and D. MERKEL: Über die Gurjunene, II: α-Gurjunen. Liebigs Ann. Chem., **617**, 129 (1958) and references cited therein.

5. (a) PALMADE, M., P. PESNELLE, J. STREITH, and G. OURISSON: L'α-Gurjunene, I: Structure et Stéréochimie. Bull. Soc. Chim. France, 1950 (1963). (b) STREITH, J., and G. OURISSON: L'α-Gurjunene, II: Quelques Réactions de l'α-Gurjunene. Bull. Soc. Chim. France, 1960 (1963).

6. PESNELLE, P., and G. OURISSON: Hydroboration of α-Gurjunene. A Rational Correlation with Cyclocolorenone. J. Organ. Chem., **30**, 1744 (1965).

7. CORBETT, R.E., and R.N. SPEDEN: The Volatile Oil of *Pseudowintera colorata*, Part II: The Structure of Cyclocolorenone. J. Chem. Soc. (London), 3710 (1958).

8. BÜCHI, G., J.M. KAUFFMAN, and J.E. LOEWENTHAL: Synthesis of 1-Epicyclocolorenone and Stereochemistry of Cyclocolorenone. J. Amer. Chem. Soc., **88**, 3403 (1966).

9. SWORDS, G., and G.L.K. HUNTER: Composition of Australian Tea Tree Oil (*Melaleuca alternifolia*). J. Agric. Food Chem., **26**, 734 (1978).

10. VAN DER GEN, A., L.M. VAN DER LINDE, and J.G. WITTEVEEN: Synthesis of Guaipyridine and Some Related Sesquiterpene Alkaloids. Rec. Trav. Chim. Pays-Bas, **91**, 1433 (1972).

11. FRIEDEL, H.D., and R. MATUSCH: New Aromadendrane Derivatives from Tolu Balsam. Helv. Chim. Acta, **70**, 1753 (1987).

12. TERHUNE, S.J., J.W. HOGG, and B.M. LAWRENCE: Essential Oils and Their Constituents, XV: β-Spathulene. A New Sesquiterpene in *Schinus molle* Oil. Phytochem., **13**, 865 (1974).

13. KINGSTON, D.G.I., M.M. RAO, T.D. SPITTLER, R.C. PETTERSEN, and D.L. CULLEN: Sesquiterpenes from *Flourensia cernua*. Phytochem., **14**, 2033 (1975).

14. BOHLMANN, F., C. ZDERO, E. HOFFMANN, P.K. MAHANTA, and W. DORNER: Naturally Occurring Terpene Derivatives, Part 166: New Diterpenes and Sesquiterpenes from South African *Helichrysum* Species. Phytochem., **17**, 1917 (1978).

15. BOHLMANN, F., M. GRENZ, J. JAKUPOVIC, R.M. KING, and H. ROBINSON: Naturally Occurring Terpene Derivatives, Part 457: Four Heliangolides and Other Sesquiterpenes from *Brasilia sickii*. Phytochem., **22**, 1213 (1983).

16. CREVOISIER, M., K.C. STEUDLE, and H.B. BUERGI: 3,3,11-Trimethyl-7-methylenetricyclo[6.3.0.0^{2,4}]undecane-5,11-diol, $C_{15}H_{24}O_2$. Acta Crystallogr., Sect. C: Cryst. Struct. Commun., **C40**, 979 (1984).

17. BOHLMANN, F., C. ZDERO, R.M. KING, H. ROBINSON: New Aromadendrene Derivatives from *Wyethia arizonica*. Planta Med., **50**, 195 (1984).

18. MARTINEZ, M., G. FLORES, A. ROMO DE VIVAR, G. REYNOLDS, and E. RODRIGUEZ: Guayulins C and D from Guayule (*Parthenium argentatum*). J. Nat. Prod., **49**, 1102 (1986).

19. (a) GOLDSBY, G., and B.A. BURKE: Sesquiterpene Lactones and a Sesquiterpene Diol from Jamaican *Ambrosia peruviana*. Phytochem., **26**, 1059 (1987). (b) SILVA, G.L., J.C. OBERTI, and W. HERZ: Sesquiterpene Lactones and Other Constituents of Argentine *Ambrosia* Species. Phytochem., **31**, 859 (1992).

20. JAKUPOVIC, J., A. SCHUSTER, F. BOHLMANN, R.M. KING, and N.S. LANDER: Sesquiterpene Lactones from *Gnephosis Species*. Phytochem., **27**, 3181 (1988).

21. JAKUPOVIC, J., L. LEHMANN, F. BOHLMANN, R.M. KING, and H. ROBINSON: Sesquiterpene Lactones and Other Constituents from *Cassinia*, *Actinobole* and *Anaxeton* Species. Phytochem., **27**, 3831 (1988).

22. JAKUPOVIC, J., A. SCHUSTER, F. BOHLMANN, R.M. KING, and L. HAEGI: Labdane

Derivatives and Other Constituents from *Waitzia acuminata*. Phytochem., **28**, 1943 (1989).

23. SAN FELICIANO, A., M. MEDARDE, M. GORDALIZA, E. DEL OLMO, and J.M. MIGUEL DEL CORRAL: Sesquiterpenoids and Phenolics of *Pulicaria paludosa*. Phytochem., **28**, 2717 (1989).

24. BOHLMANN, F., P. SINGH, and J. JAKUPOVIC: Naturally Occurring Terpene Derivatives, Part 429: Sesquiterpenes and a Dimeric Spiroketone from *Cineraria fruticulorum*. Phytochem., **21**, 2531 (1982).

25. PIZZA, C., and N. DE TOMMASI: Plant Metabolites, Part 12: Sesquiterpene Glycosides Based on the Alloaromadendrane Skeleton from *Calendula arvensis*. Phytochem., **27**, 2205 (1988).

26. JAKUPOVIC, J., M. GRENZ, F. BOHLMANN, A. RUSTAIYAN, and S. KOUSSARI: Sesquiterpene Glycosides from *Calendula persica*. Planta Med., **54**, 254 (1988).

27. DE TOMMASI, N., C. PIZZA, C. CONTI, N. ORSI, and M.L. STEIN: Structure and *in vitro* Antiviral Activity of Sesquiterpene Glycosides from *Calendula arvensis*. J. Nat. Prod., **53**, 830 (1990).

28. TAKAOKA, D., H. KAWAHARA, S. OCHI, M. HIROI, H. NOZAKI, M. NAKAYAMA, K. ISHIZAKI, K. SAKATA, and K. INA: The Structures of Sesquiterpene Glycosides from *Pittosporum tobira*. Ait. Chem. Lett., 1121 (1986).

29. BATEY, I.L., R.O. HELLYER, J.T. PINHEY: The Structure of Squamulosone, a New Sesquiterpene Ketone from *Phebalium squamulosum*. Austral. J. Chem., **24**, 2173 (1971).

30. MISKI, M., H.A. MOUBASHER, and T.J. MABRY: Sesquiterpene Aryl Esters from *Ferulago antiochia*. Phytochem., **29**, 881 (1990).

31. TRESSL, R., K.H. ENGEL, M. KOSSA, and H. KÖPPLER: Characterization of Tricyclic Sesquiterpenes in Hop (*Humulus lupulus*, var. Hersbrucker Spät). J. Agric. Food Chem., **31**, 892 (1983).

32. MAURER, B., and A. HAUSER: New Sesquiterpenoids from Clary Sage Oil (*Salvia sclarea* L.). Helv. Chim. Acta, **66**, 2223 (1983).

33. GARCIA-GRANADOS, A., and A. MOLINA: New Aromadendranic Sesquiterpenes from *Sideritis varoi* ssp. *cuatrecasasii*. Canad. J. Chem., **67**, 1288 (1989).

34. WRATTEN, S.J., and D.J. FAULKNER: Metabolites of the Red Alga *Laurencia subopposita*. J. Organ. Chem., **42**, 3343 (1977).

35. BEECHAN, C.M., C. DJERASSI, and H. EGGERT: Terpenoids, LXXIV: The Sesquiterpenes from the Soft Coral *Sinularia mayi*. Tetrahedron, **34**, 2503 (1978).

36. BRAEKMAN, J.C., D. DALOZE, R. OTTINGER, and B. TURSCH: Chemical Studies of Marine Invertebrates, XXVII: On the Absolute Configuration of Aromadendrane Sesquiterpenes from the Soft Coral *Cespitularia* aff. *subviridis*. Experientia, **33**, 993 (1977).

37. (a) BRAEKMAN, J.C., D. DALOZE, C. STOLLER, and J.P. DECLERCQ: The Configuration of Palustrol and Related Compounds. Bull. Soc. Chim. Belges, **98**, 869 (1989).
(b) CHEER, C.J., D.H. SMITH, C. DJERASSI, B. TURSCH, J.C. BRAEKMAN, and D. DALOZE: Chemical Studies of Marine Invertebrates, XVII: The Computer-Assisted Identification of (+)-Palustrol in the Marine Organism *Cespitularia* sp., aff. *subviridis*. Tetrahedron, **32**, 1807 (1976).

38. NAVES, Y.: Etudes sur les Matieres Végétales Volatiles CLXI. Présence de Lédol dans l'Huile Essentielle de Carquéja. Helv. Chim. Acta, **42**, 1996 (1959).

39. BRAEKMAN, J.C., D. DALOZE, A. DUPONT, B. TURSCH, J.P. DECLERCQ, G. GERMAIN, and M. VAN MEERSSCHE: Chemical Studies of Marine Invertebrates, XLIII: Novel Sesquiterpenes from *Clavularia inflata* and *Clavularia koellikeri* (Coelenterata, Octocorallia, Stolonifera). Tetrahedron, **37**, 179 (1981).

40. BOWDEN, B.F., J.C. COLL, L.M. ENGELHARDT, A. HEATON, and A.H. WHITE: Studies of Australian Soft Corals, XLI: Structure Determination of a New Sesquiterpene from *Xenia novae-britanniae* and an Investigation of a *Xenia* Species. Austral. J. Chem., **40**, 1483 (1987).

41. TADA, H., and F. YASUDA: Metabolites from the Marine Sponge *Epipolasis kushimotoensis*. Chem. Pharm. Bull., **33**, 1941 (1985).

42. (a) CIMINIELLO, P., E. FATTORUSSO, S. MAGNO, and L. MAYOL: New Nitrogenous Sesquiterpenes Based on Alloaromadendrane and *epi*-Eudesmane Skeletons from the Marine Sponge *Axinella cannabina*. Canad. J. Chem., **65**, 518 (1987). (b) FATTORUSSO, E., S. MAGNO, L. MAYOL, C. SANTACROCE, and D. SICA: New Sesquiterpenoids from the Sponge *Axinella cannabina*. Tetrahedron, **31**, 269 (1975). (c) FATTORUSSO, E., S. MAGNO, L. MAYOL, C. SANTACROCE, and D. SICA: Isolation and Structure of Axisonitrile-2: New Sesquiterpenoid Isonitrile from the Sponge *Axinella cannabina*. Tetrahedron, **30**, 3911 (1974).

43. CAPON, R.J., and J.K. MACLEOD: New Isothiocyanate Sesquiterpenes from the Australian Marine Sponge *Acanthella pulcherrima*. Austral. J. Chem., **41**, 979 (1988).

44. (a) BRAEKMAN, J.C., D. DALOZE, B. MOUSSIAUX, C. STOLLER, and F. DENEUBOURG: Sponge Secondary Metabolites: New Results. Pure and Appl. Chem., **61**, 509 (1989). (b) MAYOL, L., V. PICCIALLI, and D. SICA: Nitrogenous Sesquiterpenes from the Marine Sponge *Acanthella acuta*. Three New Isocyanide-Isothiocyanate Pairs. Tetrahedron, **43**, 5381 (1987). (c) BRAEKMAN, J.C., D. DALOZE, F. DENEUBOURG, J. HUYSECOM, and G. VANDEVYVER: 1-Isocyanoaromadendrane, a New Isonitrile Sesquiterpene from the Sponge *Acanthella acuta*. Bull. Soc. Chim. Belges, **96**, 539 (1987).

45. (a) FAULKNER, D.J., T.F. MOLINSKI, R.J. ANDERSEN, E.J. DUMDEI, and E. DILIP DE SILVA: Geographical Variation in Defensive Chemicals from Pacific Coast Dorid Nudibranchs and Some Related Marine Molluscs. Comp. Biochem. Physiol., C: Comp. Pharmacol. Toxicol., **97C**, 233 (1990). (b) THOMPSON, J.E., R.P. WALKER, S.J. WRATTEN, and D.J. FAULKNER: A Chemical Defense Mechanism for the Nudibranch *Cadlina luteomarginata*. Tetrahedron, **38**, 1865 (1982).

46. WU, C.L., F.F. WEY, and S.J. HSU: Sesquiterpene Hydrocarbons of the Liverwort *Scapania ornithopodioides*. Phytochem., **21**, 2659 (1982).

47. MATSUO, A., and D. TAKAOKA: Structures of New Sesquiterpenoids from the Liverwort *Mylia taylorii*. Proc. Phytochem. Soc. Eur., **29** (Bryophytes: Their Chem. Chem. Taxon.), 59 (1990).

48. ASAKAWA, Y., M. TOYOTA, and T. TAKEMOTO: Three *ent*-Secoaromadendrane-Type Sesquiterpene Hemiacetals and a Bicyclogermacrene from *Plagiochila ovalifolia* and *Plagiochila yokogurensis*. Phytochem., **19**, 2141 (1980).

49. PAKNIKAR, S.K., C.G. NAIK, N.H. ANDERSEN, and Y. OHTA: (−)-Aromadendrene (β-Diploalbicene) and (+)-*ent*-C10-Epiglobulol (Diploalbicanol) from Genus *Diplophyllum*. Indian J. Chem., **24B**, 450 (1985).

50. ASAKAWA, Y., M. TOYOTA, and T. TAKEMOTO: Sesquiterpenes from *Porella* Species. Phytochem., **17**, 457 (1978).

51. MATSUO, A., M. NAKAYAMA, S. SATO, T. NAKAMOTO, S. UTO, and S. HAYASHI: (−)-Maalioxide and (+)-Cyclocolorenone, Enantiomeric Sesquiterpenoids from the Liverwort *Plagiochila acanthophylla* subsp. *japonica*. Experientia, **30**, 321 (1974).

52. TOYOTA, M., Y. ASAKAWA, and T. TAKEMOTO: Sesquiterpenes from Japanese Liverworts. Phytochem., **20**, 2359 (1981).

53. ASAKAWA, Y., M. TOYOTA, and T. TAKEMOTO: Plagiochilide et Plagiochiline A, Secoaromadendrane-Type Sesquiterpenes de la Mousse *Plagiochila yokogurensis* (Plagiochilaceae). Tetrahedron Letters, **18**, 1553 (1978).

54. ASAKAWA, Y., M. TOYOTA, and T. TAKEMOTO: La Plagiochiline A et la Plagiochiline B, les Sesquiterpenes du Type Secoaromadendrane de la Mousse *Plagiochila hattoriana*. Phytochem., **17**, 1794 (1978).

55. MATSUO, A., K. ATSUMI, M. NAKAYAMA, and S. HAYASHI: (+)-Ovalifoliene and (−)-Hanegokedial, Two Novel Sesquiterpenoids of the *ent*-2,3-Secoalloaromadendrane Skeleton from the Liverwort *Plagiochila semidecurrens*. J. Chem. Soc. Chem. Commun., 1010 (1979).

56. MATSUO, A., H. NOZAKI, K. ATSUMI, H. KATAOKA, M. NAKAYAMA, Y. KUSHI, and S. HAYASHI: (+)-Ovalifolienalone, a Novel Sesquiterpenoid Ketone of the *ent*-2,3-Secoalloaromadendrane Group from *Plagiochila semidecurrens* (Liverwort): X-ray Crystal and Molecular Structure. J. Chem. Soc. Chem. Commun., 1012 (1979).

57. ASAKAWA, Y., M. TOYOTA, T. TAKEMOTO, and C. SUIRE: Plagiochilines C, D, E and F, Four Novel Secoaromadendrane-Type Sesquiterpene Hemiacetals from *Plagiochila asplenioides* and *Plagiochila semidecurrens*. Phytochem., **18**, 1355 (1979).

58. ASAKAWA, Y., M. TOYOTA, T. TAKEMOTO, I. KUBO, and K. NAKANISHI: Insect Antifeedant Secoaromadendrane-Type Sesquiterpenes from *Plagichila* species. Phytochem., **19**, 2147 (1980).

59. MATSUO, A., K. ATSUMI, M. NAKAYAMA, and S. HAYASHI: Structures of *ent*-2,3-Secoalloaromadendrane Sesquiterpenoids, which have Plant-Growth-Inhibitory Activity, from *Plagiochila semidecurrens* (Liverwort). J. Chem. Soc., Perkin Trans. I, 2816 (1981).

60. FUKUYAMA, Y., and Y. ASAKAWA: Neurotrophic Secoaromadendrane-Type Sesquiterpenes from the Liverwort *Plagiochila fruticosa*. Phytochem., **30**, 4061 (1991).

61. (a) MATSUO, A., S. SATO, M. NAKAYAMA, and S. HAYASHI: Taylorione, a Novel Carbon Skeletal Sesquiterpene Ketone from the Liverwort *Mylia taylorii* (Hook.) Gray. Tetrahedron Letters, 3681 (1974). (b) MATSUO, A., S. SATO, M. NAKAYAMA, and S. HAYASHI: Structure and Absolute Configuration of (−)-Taylorione, a Novel Carbon Skeletal Sesquiterpene Ketone of *ent*-1,10-Secoaromadendrane Form, from *Mylia taylorii* (Liverwort). J. Chem. Soc., Perkin Trans. 1, 2652 (1979).

62. HARRISON, L.J., and H. BECKER: A nor-Secoaromadendrane from the Liverwort *Mylia taylorri*. Phytochem., **28**, 1261 (1989).

63. (a) BENESOVÁ, V., P. SEDMERA, V. HEROUT, and F. SORM: The Structure of a Tetracyclic Sesquiterpenoic Alcohol from Liverwort *Mylia taylorii* (Hook.) Gray. Tetrahedron Letters, 2679 (1971). (b) MATSUO, A., H. NOZAKI, M. NAKAYAMA, Y. KUSHI, S. HAYASHI, N. KAMIJO, V. BENESOVÁ, and V. HEROUT: X-Ray Crystal and Molecular Structure of the *p*-Bromobenzoate of (−)-Myliol, a Novel Tetracyclic Sesquiterpene Alcohol from *Mylia taylorii* (Liverwort) Containing Two Conjugated Cyclopropane Rings. Revision of a Proposed Structure. J. Chem. Soc. Chem. Commun., 1006 (1976).

64. MATSUO, A., H. NOZAKI, M. SHIGEMORI, M. NAKAYAMA, and S. HAYASHI: (−)-Dihydromylione A, a Novel Tetracyclic Sesquiterpene Ketone Containing Two Conjugated Cyclopropane Rings, from *Mylia taylorii* (Liverwort). Experientia, **33**, 991 (1977).

65. ANDERSEN, N.H., Y. OHTA, A. MOORE, and C.W. TSENG: Anastreptene, a Commonly Encountered Sesquiterpene of Liverworts (Hepaticae). Tetrahedron, **34**, 41 (1978).

66. TAKAOKA, D., N. KOUYAMA, H. TANI, and A. MATSUO: Structures of Three Novel Dimeric Sesquiterpenoids from the Liverwort *Mylia taylorii*. J. Chem. Res. (S), 180 (1991).

67. SPÖRLE, J., H. BECKER, N.S. ALLEN, and M.P. GUPTA: Spiroterpenoids from *Plagiochila moritziana*. Phytochem., **30**, 3043 (1991).

68. WEENEN, H., M.H.H. NKUNYA, Q.A. MGANI, M.A. POSTHUMUS, R. WAIBEL, and H.

ACHENBACH: Tanzanene, a Spiro Benzopyranyl Sesquiterpene from *Uvaria tanzaniae* Verdc. J. Organ. Chem., **56**, 5865 (1991).

69. TRAUTMANN, D., B. EPE, U.E. OELBERMANN, and A. MONDON: Diterpenes from *Cneoraceae*, I: Constitution and Configuration of Cneorubines. Chem. Ber., **113**, 3848 (1980).

70. (a) AMANO, T., T. KOMIYA, M. HORI, and M. GOTO: Isolation and Characterization of Euglobals from *Eucalyptus globulus* Labill. by Preparative Reversed-Phase Liquid Chromatography. J. Chromatogr., **208**, 347 (1981). (b) KOZUKA, M., T. SAWADA, E. MIZUTA, F. KASAHARA, T. AMANO, T. KOMIYA, and M. GOTO: The Granulation-Inhibiting Principles from *Eucalyptus globulus* Labill, III: The Structures of Euglobal-III, -IVb and -VII. Chem. Pharm. Bull., **30**, 1964 (1982).

71. NISHIZAWA, M., M. EMURA, Y. KAN, H. YAMADA, K. OGAWA, and N. HAMANAKA: Macrocarpals: HIV-Reverse Transcriptase Inhibitors of *Eucalyptus globulus*. Tetrahedron Letters, **33**, 2983 (1992).

72. MURATA, M., Y. YAMAKOSHI, S. HOMMA, K. AIDA, K. HORI, and Y. OHASHI: Macrocarpal A, a Novel Antibacterial Compound from *Eucalyptus macrocarpa*. Agric. Biol. Chem., **54**, 3221 (1990).

73. YAMAKOSHI, Y., M. MURATA, A. SHIMIZU, and S. HOMMA: Isolation and Characterization of Macrocarpals B–G, Antibacterial Compounds from *Eucalyptus macrocarpa*. Biosci. Biotechnol. Biochem., **56**, 1570 (1992).

74. DOLEJS, L., V. HEROUT, O. MOTL, F. SORM, and M. SOUCEK: Epimeric Aromadendrenes: Stereoisomerism of Ledol, Viridiflorol and Globulol. Chem. and Ind., 566 (1959) and references cited.

75. SEMMLER, F.W., and E. TOBIAS: Zur Kenntnis der Bestandteile Ätherischer Öle (über Eudesmol und seine Derivate; Notiz über Globulol). Ber. dtsch. chem. Ges., **46**, 2030 (1913).

76. KIR'YALOV, N.P.: Principal Components of the Essential Oil of *Ledum palustre*. Doklady Akad. Nauk (SSSR), **61**, 305 (1948) (Chem. Abstr., **43**, 1155e (1949)).

77. JONES, T.G.H., and W.L. HAENKE: Essential Oils from the Queensland Flora, XI: *Melaleuca viridiflora*, Part II. Proc. Roy. Soc. Queensland, **49**, 95 (1938).

78. BOWYER, R.C., and P.R. JEFFERIES: Structure of Spathulenol. Chem. and Ind., 1245 (1963).

79. JENNISKENS, L.H.D., J.B.P.A. WIJNBERG, and AE. DE GROOT: Base-Induced and -Directed Elimination and Rearrangement of Perhydronaphthalene-1,4-diol Monosulfonate Esters. Total Synthesis of ($\pm$)-Alloaromadendrane-4β,10α-diol and ($\pm$)-Alloaromadendrane-4α,10α-diol. J. Organ. Chem., **56**, 6585 (1991).

80. (a) CANE, D.E.: In: Biosynthesis of Isoprenoid Compounds (PORTER, J.W., and S.L. SPURGEON, eds.), Vol. 1, p. 283, John Wiley & Sons, New York, 1981. (b) MANITTO, P.: Biosynthesis of Natural Products, p. 238, Ellis Horwood Ltd., Chichester, 1981. (c) RUZICKA, L.: History of the Isoprene Rule. Proc. Chem. Soc. (London), 341 (1959). (d) HENDRICKSON, J.B.: Stereochemical Implications in Sesquiterpene Biogenesis. Tetrahedron, **7**, 82 (1959). (e) PARKER, J.W., J.S. ROBERTS, and R. RAMAGE: Sesquiterpene Biogenesis. Quart. Rev. (Chem. Soc. London), **31**, 331 (1967).

81. DEVON, T.K., and A. SCOTT: Handbook of Naturally Occurring Compounds, Vol. II: Terpenes, p. 56, Academic Press, New York, 1972.

82. NISHIMURA, K.: A New Sesquiterpene, Bicyclogermacrene. Tetrahedron Letters, 3097 (1969).

83. MATSUO, A., H. NOZAKI, N. KUBOTA, S. UTO, and M. NAKAYAMA: Structures and Conformations of (−)-Isobicyclogermacrenal and (−)-Lepidozenal, Two Key Sesquiterpenoids of the *cis*- and *trans*-10,3-Bicyclic Ring Systems, from the Liverwort

Lepidozia vitrea: X-Ray Crystal Structure Analysis of the Hydroxy Derivative of (−)-Isobicyclogermacrenal. J. Chem. Soc., Perkin Trans. 1, 203 (1984).

84. For example, see refs. no. *31, 81, 92c*.

85. For example, see refs, no. *48, 50, 52*, and *57*.

86. Nishimura, K., I. Horibe, and K. Tori: Conformations of 10-Membered Rings in Bicyclogermacrene and Isobicyclogermacrene. Tetrahedron, **29**, 271 (1973).

87. (a) Marshall, J.A., and W.F. Huffman: A New Synthetic Approach to Hydroazulenes. J. Amer. Chem. Soc., **92**, 6358 (1970). (b) Concannon, P.W., and J. Ciabattoni: Peroxy Acid Oxidation of Cycloalkynes and the Decomposition of 2-Diazocycloalkanones. J. Amer. Chem. Soc., **95**, 3284 (1973). (c) See ref. *97*.

88. (a) Rücker, G., R. Mayer, H. Wiedenfeld, B.S. Chung, and A. Güllmann: (+)-Isobicyclogermacrenal from *Aristolochia manshuriensis*. Phytochem., **26**, 1529 (1987). (b) Paliwal, M.K., I.R. Siddiqui, S. Singh, H.P. Tiwari: Phytochemical Investigation of *Asterella angusta*. J. Indian Chem. Soc., **68**, 533 (1991).

89. Garson, M.J.: Biosynthesis of the Novel Diterpene Isonitrile Diisocyanoadociane by a Marine Sponge of the *Amphimedon* Genus: Incorporation Studies with Sodium [^{14}C]Cyanide and Sodium [2-^{14}C]Acetate. J. Chem. Soc., Chem. Commun., 35 (1986).

90. Hagadone, M.R., P.J. Scheuer, and A. Holm: On the Origin of the Isocyano Function in Marine Sponges. J. Amer. Chem. Soc., **106**, 2447 (1984).

91. For example, see refs. no. *4, 9, 11, 32*, and *123*.

92. Katsiotis, S.T., C.R. Langezaal, J.J.C. Scheffer, and R. Verpoorte: Comparative Study of the Essential Oils from Hops of Various *Humulus lupulus* L. Cultivars. Flavour Fragrance J., **4**, 187 (1989).

93. For example see (a) Chalchat, J.C., R.P. Garry, A. Michet, and L. Peyron: Chemical Composition of Natural and Empyreumatic Oils and Extracts from *Juniperus oxycedrus* and *Juniperus phoenicea* Wood. J. Essent. Oil Res., **2**, 231 (1990). (b) Ji, X., Q. Pu, H.M. Garraffo, and L.K. Pannell: The Essential Oil of the Leaves of *Psidium guajava* L. J. Essent. Oil Res., **3**, 187 (1991). (c) Aalbersberg, W.G.L., and Y. Singh: Essential Oils from Two Medicinal Plants of Fiji: *Dysoxylum richii* (A. Gray) C.D.C. Fruit and *Synedrella nodiflora* (L.) Gaertn. Leaves. Flavour Fragrance J., **6**, 125 (1991). (d) De Bernardi, M., G. Vidari, P. Vida-Finzi, S. Abdo, G. Marinoni, and G. Mellerio: Medicinal Plant Metabolites, III: GC-MS Analysis of the Essential Oil of *Lasiocephalus ovatus*. Rev. Latinoam. Quim., **21**, 97 (1990). (e) Ji, X., Q. Pu, H.M. Garraffo, and L.K. Pannell: The Essential Oil of the Leaves of *Callistemon rigidus* R. Br. J. Essent. Oil Res., **3**, 465 (1991). (f) Onayade, O.A., J.J.C. Scheffer, and A. Baerheim Svendsen: Polynuclear Aromatic Compounds and Other Constituents of the Herb Essential Oil of *Salvia coccinea* Juss. ex Murr. Flavour Fragrance J., **6**, 281 (1991).

94. For example, see (a) Ekundayo, O., I. Laakso, M. Holopainen, R. Hiltunen, B. Oguntimein, and V. Kauppinen: The Chemical Composition and Antimicrobial Activity of the Leaf Oil of *Vitex agnus-castus* L. J. Essent. Oil Res., **2**, 115 (1990). (b) Chalchat, J.C., R.Ph. Garry, A. Michet, P. Bastide, and R. Malhuret: Chemical Composition/Antimicrobial Activity Correlation, IV: Comparison of the Activity of Natural and Oxygenated Essential Oils Against Six Strains. Plant. Med. Phytother., **23**, 305 (1989). (c) Iwu, M.M., C.O. Ezeugwu, C.O. Okunji, D.R. Sanson, and M.S. Tempesta: Antimicrobial Activity and Terpenoids of the Essential Oil of *Hyptis suaveolens*. Int. J. Crude Drug Res., **28**, 73 (1990). (d) Urzua, A.M., and R.A. Rodriguez: Germination-Inhibiting Terpenes from the Roots of *Aristolochia chilensis*. Bol. Soc. Chil. Quim., **37**, 183 (1992).

95. Tkhu, D.Kh., V.I. Roshchin, O.N. Malysheva, and V.A. Solov'ev: Fungicidal

Activity of an Extract from Wood of *Taubauma gioi*. Khim. Drev., 103 (1987) (Chem. Abstr., **106**, 99457w (1988)).

96. HUBERT, T.D., and D.F. WIEMER: Ant-Repellent Terpenoids from *Melampodium divaricatum*. Phytochem., **24**, 1197 (1985).

97. GIJSEN, H.J.M., J.B.P.A. WIJNBERG, G.A. STORK, AE. DE GROOT, M.A. DE WAARD, and J.G.M. VAN NISTELROOY: The Synthesis of Mono- and Dihydroxy Aromadendrane Sesquiterpenes, Starting from Natural (+)-Aromadendrene, III. Tetrahedron, **48**, 2465 (1992).

98. JACYNO, J.M., N. MONTEMURRO, A.D. BATES, and H.G. CUTLER: Phytotoxic and Antimicrobial Properties of Cyclocolorenone from *Magnolia grandiflora* L. J. Agric. Food Chem., **39**, 1166 (1991).

99. BOLTE, M.L., J. BOWERS, W.D. CROW, D.M. PATON, A. SAKURAI, N. TAKAHASHI, M. UJIIE, and S. YOSHIDA: Germination Inhibitor from *Eucalyptus pulverulenta*. Agric. Biol. Chem., **48**, 373 (1984).

100. MESSER, A., K. MCCORMICK, SUNJAYA, H.H. HAGEDORN, F. TUMBEL, and J. MEINWALD: Defensive Role of Tropical Tree Resins: Antitermitic Sesquiterpenes from Southeast Asian *Dipterocarpaceae*. J. Chem. Ecol., **16**, 3333 (1990).

101. HARADA, A., K. SAKATA, and K. INA: A New Screening Method for Antifouling Substances Using the Blue Mussel *Mytilus edulis* L. Agric. Biol. Chem., **48**, 641 (1984).

102. THOMPSON, J.E., R.P. WALKER, and D.J. FAULKNER: Screening and Bioassays for Biologically Active Substances from Forty Marine Sponge Species from San Diego, California, USA, Mar. Biol. (Berlin), **88**, 11 (1985).

103. ASAKAWA, Y.: Phytochemistry of *Hepaticae*: Isolation of Biologically Active Aromatic Compounds and Terpenoids. Rev. Latinoamer. Quim., **14**, 109 (1984).

104. LIST, P.H., and L. HÖRHAMMER: Hagers Handbuch der Pharmazeutischen Praxis, 4th Ed., Vol. 5, pp. 479–480, Springer-Verlag, Berlin-Heidelberg-New York, 1976.

105. FUKUYAMA, Y., and Y. FUKUYAMA: Nerve Cell Degeneration Reparation Agents Containing Secoaromadendrane-Type Sesquiterpenes and/or Plagiochilide from *Plagiochila fruticosa*. Jpn. Kokai Tokyo Koho (1991).

106. MURATA, M., Y. YAMAKOSHI, S. HOMMA, K. ARAI, and Y. NAKAMURA: Macrocarpals, Antibacterial Compounds from *Eucalyptus*, Inhibit Aldose Reductase. Biosci., Biotechnol., Biochem., **56**, 2062 (1992).

107. HO, T.L.: Carbocycle Construction in Terpene Synthesis, pp. 589–594, VCH Publishers, New York, 1988.

108. SURBURG, H., and A. MONDON: Synthesis of (−)-Spathulenol. Chem. Ber., **114**, 118 (1981).

109. BARTON, D.H.R., P. DE MAYO, and M. SHAFIQ: Photochemical Transformations, Part I: Some Preliminary Investigations. J. Chem. Soc. (London), 929 (1957).

110. STREITH, J., and A. BLIND: Stereospecific Photochemical Synthesis of Some Aromadendrane Derivatives. Bull. Soc. Chim. Fr., 2133 (1968).

111. CAINE, D., and P.F. INGWALSON: The Influence of Substituents on the Photochemical Behavior of Cross-Conjugated Cyclohexadienones. A Facile Total Synthesis of (−)-Cyclocolorenone. J. Organ. Chem., **37**, 3751 (1972).

112. CAINE, D., and J.T. GUPTON III: Photochemical Rearrangements of Cross-Conjugated Cyclohexadienones. Application to the Synthesis of (−)-4-Epiglobulol and (+)-4-Epiaromadendrene. J. Organ. Chem., **40**, 809 (1975).

113. ROMBERGER, M.L.: Allylidenecyclopropanes: Their Synthesis via α-Lithiosilanes and Their Synthetic Application. Synthesis of α-Bulnesol and Studies Directed Towards the Synthesis of (+)-Ledene. Dissertation (1989). Avail. Univ. Microfilms Int., Order No. DA9010023. From: Diss. Abstr. Int. B, **50**(11), 5076 (1990).

114. Narang, S.A., and P.C. Dutta: Synthetical Studies of Terpenoids, Part VIII: Synthesis of an Isomer of (±)-Cyclocolorenone. J. Chem. Soc. (London), 1119 (1964).

115. Marshall, J.A., and J.A. Ruth: Synthesis of Racemic Globulol via Solvolysis-Cyclization of a 2,7-Cyclodecadien-1-ol Derivative. J. Organ. Chem., 39, 1971 (1974).

116. Nakayama, M., S. Ohira, S. Shinke, Y. Matsushita, A. Matsuo, and S. Hayashi: Synthesis of (−)-Taylorione, a Sesquiterpene Ketone of ent-1,10-Secoaromadendrane Skeleton. Chem. Lett., 1245 (1979).

117. Pattenden, G., and D. Whybrow: Synthetic Photochemistry. A Synthesis of the Carbon Skeleton Found in Taylorione from Mylia taylorii, Using the Di-π-methane Rearrangement. J. Chem. Soc., Perkin Trans. I, 1046 (1981).

118. Taylor, M.D., G. Minaskanian, K.N. Winzenberg, P. Santone, and A.B. Smith III: Preparation, Stereochemistry, and Nuclear Magnetic Resonance Spectroscopy of 4-Hydroxy(acetoxy)bicyclo[5.1.0]octanes. Synthesis of (−)- and (±)-8,8-Dimethyl-bicyclo[5.1.0]oct-2-en-4-one. J. Organ. Chem., 47, 3960 (1982).

119. Taylor, M.D., and A.B. Smith III: Total Synthesis of (+)-Hanegokedial. Tetrahedron Letters, 24, 1867 (1983).

120. Büchi, G., S.W. Chow, T. Matsuura, T.L. Popper, H.H. Rennhard, and M. Schach von Wittenau: Terpenes, XII: The Constitutions of Aromadendrene, Globulol, Ledol and Viridiflorol. Tetrahedron Letters, 14 (1959).

121. Graham, B.A., P.R. Jefferies, G.J.H. Melrose, K.J.L. Thieberg, and D.E. White: The Stereochemistry of Aromadendrene, Globulol, and Ledol. Austral. J. Chem., 13, 372 (1960).

122. Rienäcker, R., and J. Graefe: Catalytic Conversion of Sesquiterpene Hydrocarbons on Alkali Metal/Aluminium Oxide Contacts. Angew. Chem., 97, 348 (1985).

123. Gijsen, H.J.M., and G.A. Stork: Unpublished results.

124. Van Lier, F.P., T.G.M. Hesp, L.M. Van der Linde, and A.J.A. Van der Weerdt: First Preparation of (+)-Spathulenol. Regio- and Stereoselective Oxidation of (+)-Aromadendrene with Ozone. Tetrahedron Letters, 26, 2109 (1985).

125. Gijsen, H.J.M., K. Kanai, G.A. Stork, J.B.P.A. Wijnberg, R.V.A. Orru, C.G.J.M. Seelen, S.M. Van der Kerk, and Ae. de Groot: The Conversion of Natural (+)-Aromadendrene into Chiral Synthons, I. Tetrahedron, 46, 7237 (1990).

126. Gijsen, H.J.M., J.B.P.A. Wijnberg, C. Van Ravenswaay, and Ae. de Groot: Rearrangement Reactions of Aromadendrane Derivatives. The Synthesis of (+)-Maaliol, Starting from Natural (+)-Aromadendrene, IV. Tetrahedron, 50, 4733 (1994).

127. Gijsen, H.J.M., J.B.P.A. Wijnberg, G.A. Stork, and Ae. de Groot: The Synthesis of (−)-Kessane, Starting from Natural (+)-Aromadendrene, II. Tetrahedron, 47, 4409 (1991).

128. Gijsen, H.J.M., J.B.P.A. Wijnberg, and Ae. de Groot: Thermal Rearrangement of Bicyclogermacrane-1,8-dione. The Synthesis of Humulenedione and (−)-Cubenol, Starting from Natural (+)-Aromadendrene, V. Tetrahedron, 50, 4745 (1994).

129. Abraham, W.R., K. Kieslich, B. Stumpf, and L. Ernst: Microbial Oxidation of Tricyclic Sesquiterpenoids Containing a Dimethylcyclopropane Ring. Phytochem., 31, 3749 (1992).

130. Abraham, W.R., L. Ernst, B. Stumpf, and H.A. Arfmann: Microbial Hydroxylations of Bicyclic and Tricyclic Sesquiterpenes. J. Essent. Oil Res. 1, 19 (1989).

131. Asakawa, Y., T. Ishida, M. Toyota, and T. Takemoto: Terpenoid Biotransformation in Mammals, IV: Biotransformation of (+)-Longifolene, (−)-Caryophyllene, (−)-Caryophyllene oxide, (−)-Cyclocolorenone, (+)-Nootkatone, (−)-Elemol, (−)-Abietic Acid and (+)-Dehydroabietic Acid in Rabbits. Xenobiotica, 16, 753 (1986).

132. EHRET, C., and G. OURISSON: Le γ-Gurjunene, Structure et Configuration. Isomerisation de L'α-Gurjunene. Tetrahedron, **25**, 1785 (1969).

133. FRIEDEL, H.D., and R. MATUSCH: Isolation and Structure Elucidation of Epimeric 1(5),6-Guaiadienes from Tolu Balsam. Helv. Chim. Acta, **70**, 1616 (1987).

134. MEHTA, G., and B.P. SINGH: Terpenes and Related Systems, 16: Fate of Representative Bicyclic Sesquiterpenes in Strong Acid Medium. A General Rearrangement of Hydroazulene Sesquiterpenes to Decalin Types. J. Organ. Chem., **42**, 632 (1977).

135. RICHARDSON, D.P., A.C. MESSER, B.A. NEWTON, and N.I. LINDEMAN: Identification and Preparation of Antiinsectan Dienols from *Dipterocarpus kerrii* Tree Resins. J. Chem. Ecol., **17**, 663 (1991).

136. TREIBS, W., and H.M. BARCHET: Über bi- und polycyclische Azulene, IV: Das Aromadendren, sein chemischer Bau und seine Überführung in 5-Azulene. Liebigs Ann. Chem., **566**, 89 (1950).

137. TAKESHITA, H., M. HIRAMA, and S. ITO: Conversion of Gurjunene to 10-Epizierone. Stereochemistry of Zierone. Tetrahedron Letters, 1775 (1972).

(Received February 5, 1994)

Author Index

Page numbers printed in *italics* refer to References

Subject Index

Fortschritte der Chemie organischer Naturstoffe

Progress in the Chemistry of Organic Natural Products

Volume 63

1994. VII, 216 pages. Cloth DM 220,–, öS 1540,–
Subscription price: Cloth DM 198,–, öS 1386,–
ISBN 3-211-82443-X

Contents: L. Rodríguez-Hahn, B. Esquivel, and J. Cárdenas: Clerodane Diterpenes in Labiatae. • Anil B. Ray and Mohini Gupta: Withasteroids, a Growing Group of Naturally Occuring Steroidal Lactones. • Author Index. • Subject Index.

Volume 62

1993. 52 figures. VII, 330 pages. Cloth DM 280,–, öS 1960,–
Subscription price: Cloth DM 252,–, öS 1764,–
ISBN 3-211-82402-2

Contents: S.V. Bhat: Forskolin and Congeners. • L. Minale, R. Ricci, and F. Zollo: Steroidal Oligoglycosides and Polyhydroxysteroids from Echinoderms. • Author Index. • Subject Index.

Volume 61

1993. 4 figures. IX, 206 pages. Cloth DM 220,–, öS 1540,–
Subscription price: DM 198,–, öS 1386,–
ISBN 3-211-82388-3

Contents: D.G.I. Kingston, A.A. Molinero, and J.M. Rimoldi: The Taxane Diterpenoids. • Author Index. • Subject Index.

Springer-Verlag Wien New York

Sachsenplatz 4–6, P.O.Box 89, A-1201 Wien · 175 Fifth Avenue, New York, NY 10010, USA
Heidelberger Platz 3, D-14197 Berlin · 3-13, Hongo 3-chome, Bunkyo-ku, Tokyo 113, Japan

Volume 60

1992. 59 figures. VIII, 243 pages.
Cloth DM 184,–, öS 1290,–
Subsciption price: Cloth DM 165,60, öS 1161,–*)
ISBN 3-211-82374-3

Contents: I. Wahlberg and A.-M. Eklund: Cyclized Cembranoids of Natural Occurrence. • M. Petitou and C.A.A. van Boeckel: Chemical Synthesis of Heparin Fragments and Analogues. • Author Index. • Subject Index. • General Index Vols. 21–60.

*) **Special Offer:** *Special reduced price (20% reduction) for the complete Series Vols. 1–60 incl. Cumulative Index to Vols. 1–20.*

Volume 59

1992. 1 figure. IX, 328 pages.
Cloth DM 260,–, öS 1820,–
Subscription price: Cloth DM 234,–, öS 1638,–
ISBN 3-211-82278-X

Contents: Shin-Ichi Hatanaka: Amino Acids from Mushrooms. • I. Wahlberg and A.-M. Eklund: Cembranoids, Pseudopteranoids, and Cubitanoids of Natural Occurence. • Author Index. • Subject Index.

Volume 58

1991. 64 figures. VII, 343 pages.
Cloth DM 280,–, öS 1960,–
Subscription price: Cloth DM 252,–, öS 1764,–
ISBN 3-211-82265-8

Contents: J. A. Robinson: Chemical and Biochemical Aspects of Polyether-Ionophore Antibiotic Biosynthesis. • R. D. H. Murray: Naturally Occuring Plant Coumarins. • Formula Index. • Trivial Name Index.

Springer-Verlag Wien New York

Sachsenplatz 4–6, P.O.Box 89, A-1201 Wien · 175 Fifth Avenue, New York, NY 10010, USA
Heidelberger Platz 3, D-14197 Berlin · 3-13, Hongo 3-chome, Bunkyo-ku, Tokyo 113, Japan

Springer-Verlag
and the Environment

WE AT SPRINGER-VERLAG FIRMLY BELIEVE THAT AN international science publisher has a special obligation to the environment, and our corporate policies consistently reflect this conviction.

WE ALSO EXPECT OUR BUSINESS PARTNERS – PRINTERS, paper mills, packaging manufacturers, etc. – to commit themselves to using environmentally friendly materials and production processes.

THE PAPER IN THIS BOOK IS MADE FROM NO-CHLORINE pulp and is acid free, in conformance with international standards for paper permanency.